T0271370

# Wireless Local Area Networks

# Wireless Local Area Networks
## The New Wireless Revolution

*Edited by*

**Benny Bing**
*Georgia Institute of Technology*

WILEY-INTERSCIENCE

A JOHN WILEY & SONS, INC., PUBLICATION

*Library of Congress Cataloging-in-Publication Data is available.*

ISBN 0-471-22474-X

10 9 8 7 6 5 4 3 2 1

# Contents

               James Chen
               *Atheros Communications, Inc.*

**Chapter 4    Migration Strategies for IEEE 802.11 Wireless LANs 91**
*Proxim, Inc.*

**Chapter 5    5-GHz Radio Spectrum Regulations                    97**
Teik-Kheong Tan
*3Com Corporation, Inc.*

**Chapter 6    Quality of Service and Multimedia Support in**
**802.11 Standards**                                    **105**
Gregory Parks
*Cirrus Logic, Inc.*

**Chapter 7    Overview of Wireless LAN Security**         **115**
*Cisco Systems*

# Contributors

WildPackets, Inc.
2540 Camino Diablo
Walnut Creek, CA 94596, USA

Chris Heegard
Texas Instruments
Home and Wireless Networking
141 Stony Circle, Suite 130
Santa Rosa, CA 95401, USA

James Chen
Atheros Communications, Inc.
529 Almanor Ave.
Sunnyvale, CA 94085-3512, USA

Proxim, Inc.
510 DeGuigne Dr.
Sunnyvale, CA 94085, USA

Teik-Kheong Tan
3Com
5400 Bayfront Plaza
Santa Clara, CA 95052-8145, USA

Greg Parks
Cirrus Logic, Inc.
Wireless Networking Division
5175 Hillsdale Circle
El Dorado Hills, CA 95762, USA

Cisco Systems, Inc.
170 West Tasman Dr.
San Jose, CA 95134-1706, USA

Dorothy Stanley
Agere Systems
2000 North Naperville Rd.
Naperville, IL 60566, USA


Colubris Networks
440 Armand-Frappier, Suite 200
Laval, Quebec H7V 4B4, Canada

Philippe Laine
Alcatel Network Strategy Group
54 rue La Boëtie
75411 Paris Cedex 08, France

Jouni Mikkonen
Nokia Mobile Phones
P.O. Box 88
FIN-33721 Tampere, Finland

Sandeep Singhal
ReefEdge, Inc.
2 Executive Drive, Suite 600
Fort Lee, NJ 07024, USA

Mike Sheppard
Ericsson Wireless Communications, Inc.
6455 Lusk Blvd.
San Diego, CA 92121-2779, USA

Stephen J. Shellhammer
Symbol Technologies, Inc.
One Symbol Plaza, Mail Stop B-2
Holtsville, NY 11742-1300, USA

Kazimierz Siwiak
Time Domain Corporation
Cummings Research Park
7057 Old Madison Pike
Huntsville, AL 35806, USA

# Foreword

Wireless local area networks do indeed represent the new wireless revolution. This book presents the case for why the technology has emerged, how it works, what the issues are, and where it is going. It presents the thinking of a collection of experts in each of many individual areas and so comes across as a detailed presentation covering all aspects of wireless LANs from the underlying technology all the way to the business case for deployment.

We are witnessing a grassroots movement with wireless LANs that has the potential to compete with wireless data technologies being offered to us by the telecommunication carriers. This is not the first time we have seen such a struggle: We saw it in the Ethernet vs. ATM wars, we saw it in the packet switching vs. circuit switching wars, and we have seen it in other arenas as well. In the case of wireless LANs, we are not likely to see a "winner-take-all" outcome because there seems to be a place for the cellular systems solutions as well.

The origins of the vigorous Ethernet technology come from the early studies that took place in the mid-1970s when we analyzed the behavior of Carrier Sense Multiple Access (CSMA) in a radio environment. When CSMA was modified to apply to a wired environment with the addition of Collision Detection (which does not work well with wireless links) to produce CSMA/CD, Ethernet was born. Of course, the Ethernet technology has evolved well beyond that original protocol, and we have seen Ethernet emerge in today's wireless link technology where, once again, CD cannot be used directly. So we have indeed come full circle where again CSMA in its basic form provides the underlying technology for wireless LANs. MAC layer issues such as this are discussed in this book.

The impact of wireless LANs has the promise of revolutionizing the network landscape. The WiFi movement, as defined by the Wireless Ethernet Compatibility Alliance (WECA) is actively defining best-practices technology for use in the industry. As we see WiFi roll out in its many forms as described in this book, we will also see the growth of nomadic computing in which access to the Internet can be obtained easily by any user in any location with any device. Wireless LANs are but one, albeit important, technology to bring about the support of nomadic computing. There are additional needed technologies; for example, we also need access control, configuration management, subscriber management, billing, provisioning of services, policy-based traffic shaping, and more. With this full set of technologies, we will see a very rapid rise in the deployment of nomadic computing capability and the consequent impact on networking in general.

The material in this book is timely indeed and represents the latest word on wireless LAN technology.

Leonard Kleinrock
Professor of Computer Science, UCLA
Chairman, Nomadix Inc.

# Preface

Wireless local area networks (LANs) are becoming ubiquitous and increasingly relied upon. From airport lounges and hotel meeting rooms to cafés and restaurants across the globe, wireless LANs are being built for mobile professionals to stay connected to the Internet. The fact that megastores such as Starbucks and Sears are deploying wireless LANs shows how prevalent the technology has now become. In the year 2000, the sale of IEEE 802.11b wireless LAN adapters have increased dramatically from about 5,000 to 70,000 units per month. Currently, nearly a million 802.11b adaptors are being sold per month and newer versions of notebook computers are starting to have such adapters integrated into them. These developments rival the popularity of wired Ethernet networks.

Was such immense success anticipated? Just like the emergence of E-mail and the Web browser, the answer is an emphatic "No", simply because in the last few years many companies have invested heavily on wireless technologies that promised to emulate the mass-market success of voice-based second-generation (2G) cell phones. These include communication over low-orbit satellites and third-generation (3G) cell phones, both of which have fallen short of expectations. Why then have wireless LANs taken off so rapidly? One important reason is that such networks combine the power of wireless access with mobile computing, delivering high data rates on unlicensed radio spectrum for office, home, and public networks. Although interference is an issue, unlicensed frequency bands provide a far more significant advantage in that they remove exorbitant overheads associated with acquiring radio spectrum, thereby lowering the barrier for new entrants. This in turn drives down product costs because of increased competition. This advantage, coupled with improved data rates and the proliferation of cheaper but more powerful personal devices, has accelerated the demand for wireless LANs tremendously in the last few years. The increasing attractiveness of public access networks that involve the integration of high-speed wireless LANs and cellular networks under a unified billing/identification system is poised to launch another rich avenue of growth in the future. Wireless LANs need not transfer purely data traffic. They can also support packetized voice and video transmission. People today are spending huge amounts of money, even from office to office, calling by cellular phones. With a wireless LAN infrastructure, it costs them a fraction of what it cost them to use cell phones or any other equipment.

This book discusses the latest developments on the wireless LAN standards as well as emerging enhancements and applications. A unique feature of this book is that almost all chapters are contributed by major players in the wireless LAN industry. Although some chapters are written specifically for this book, others are obtained from presentations delivered at the 2001 International Conference on Wireless LANs and Home Networks as well as company white papers. They are primarily selected on the basis of conciseness and clarity. These selection criteria cut down unnecessary information, reduce the time needed to learn new concepts, and increase the accessibility of the book to readers with differing backgrounds. I believe the reduction in excess information comes as a welcome relief in an Internet age in which we are all deluged with excessive information. As you will discover, the chapters deal with a broad range of topics, covering almost every aspect of wireless LAN design including important insights and updates on standards, technology, and applications. These chapters can be further categorized under standards (Chapters 1–6), network security (Chapters 7–9), public access (Chapters 10–12), personal area networks (Chapters 13 and 14), and future technology (Chapter 15). Because each chapter is self-contained some overlap in material is inevitable among different chapters, although this is minimized. A positive side of such overlap, however, is that each affected chapter emphasizes a different perspective of the same issues.

There are three major IEEE wireless LAN standards (802.11a, 802.11b, and 802.11g) operating on different frequency bands (2.4 and 5 GHz), and it is unclear which standard will eventually prevail. The 802.11g draft standard provides data rates of up to 54 Mbit/s using OFDM in the 2.4 GHz band. It is expected to be approved in early 2003 and is backward-compatible with 802.11b. 802.11b currently enjoys an overwhelming majority of all wireless LAN installations but suffers from interference in the congested 2.4-GHz band and a lower data rate (compared with 802.11a). The main source of interference stems from high-power microwave ovens, which reside in practically every office and home. With new wireless personal devices involving Bluetooth starting to gain traction commercially, the interference in the 2.4-GHz is poised to become intolerable. 802.11a is touted to be a natural migration from 802.11b by offering higher data rates and a cleaner 5-GHz band. However, the higher frequency of operation also means that for the same data rate, the operational range is shorter compared with 2.4-GHz systems. Range is a critical factor because a wireless LAN with restricted range performance implies the need for more wireless access points, which are significantly more expensive than portable wireless adapter cards. Clearly, a network with many access points will increase deployment costs. A solution to this constraint is to increase the operational distance of 802.11a systems at the expense of data rate performance. The question is how much data rate can be sacrificed before a breakeven point is reached where the range performance of 802.11a becomes comparable with 802.11b. This interesting trade-off issue is examined by some chapters of this book. Another drawback of 802.11a systems is that such systems currently cannot operate in western Europe because the 5-GHz band has been reserved for another wireless LAN standard—HiperLAN2. Although HiperLAN2 employs the same physical multiplexing scheme as 802.11a, some of the parameters are different. In addition, the access methods are completely incompatible. Unless these differences are resolved quickly, the popularity of 802.11a in Europe can be severely curtailed while the 802.11b market gains even more momentum.

For the beginner or the uninitiated, the IEEE 802.11 standards can be daunting documents to read. The first chapter relieves this pain considerably by providing a concise treatment of wireless LANs and the IEEE 802.11b standard. The key to understanding the standard is to be clear about the organizational structure and the reasons behind this structure. Chapter 1 achieves this goal by emphasizing the important elements of the standard and how they relate to one another. The tables summarizing the various protocol functions are especially useful. This chapter also examines the application of wireless LAN analyzers in solving network problems, including enhancing security, simplifying network management, and optimizing network performance. Although not discussed in the chapter, such analyzers play an even more important role in interoperability testing and are a sound investment for companies with no prior experience in developing 802.11b network adapter cards.

The next few chapters provide a more in-depth discussion of the IEEE wireless LAN standards. Chapter 2 begins with a description of the evolution of the 802.11b standard. It focuses on the media access control layer and then presents details of the physical layer transmission methods, including coding descriptions and performance evaluations. The coding schemes employed by 802.11b are nontrivial and require in-depth analysis. To this end, extensive explanations and examples are included in this chapter to illustrate difficult coding concepts such as Barker, Complementary Code Keying (CCK), and Packet Binary Convolutional Coding (PBCC) coding. The Barker coding scheme was employed in the first version of the 802.11 standard (issued in 1997), whereas the CCK coding scheme was introduced in the current 802.11b standard (issued in 1999). The PBCC coding method is an option in the 802.11b standard as well as the new 2.4-GHz 802.11g draft standard. The later sections of Chapter 2 explain the role and limitations of spread spectrum communications when applied to 802.11b. The discussion of how the FCC modified the theoretical rules of spread spectrum transmission to accommodate practical implementations is particularly interesting. Finally, the chapter concludes with a detailed coverage of range versus rate performance of several physical layer standards in the 802.11g draft standard.

Chapter 3 by James Chen of Atheros Communications presents an introduction to 802.11a systems, with emphasis on the fundamentals of Orthogonal Frequency Division Multiplexing (OFDM). OFDM is a multiplexing technique that forms the basis of the current 5-GHz 802.11a standard and possibly higher-speed enhancements of the 2.4-GHz 802.11b standard in the future. The method involves combining many radio carriers, each transporting a portion of the information contained in a data packet. It is not a modulation scheme as some people seem to imply and does not exclude the use of other modulation schemes. In fact, the 802.11a standard defines four different modulation schemes to be used in conjunction with OFDM. OFDM is only recently becoming popular because integrated chips that can perform high speed Fast Fourier Transform (FFT) in real time have become economical. The FFT algorithm generates different carriers while ensuring that the peak of each carrier fits into the nulls of all other carriers, thus giving rise to orthogonal transmission. OFDM avoids multipath problems caused by signal reflections, which is a major problem in indoor wireless communications. Each carrier transports information at a rate slow enough so that any delayed copies due to reflections affect only a small fraction of a data symbol period. Multiple carriers are sent at the same time, and these carriers are combined at the receiver, thereby achieving the equivalent data rate of a single high-speed carrier. Whereas Chapter 2 focuses on analytical and simulation results, Chapter 3 presents measured 5-GHz 802.11a range versus rate performance data. The rate performance is further refined in terms of for raw wireless bit rate (data link rate) and usable throughput. These results are then used to calculate the overall 802.11a system capacity in a multicell radio environment. It is claimed that 802.11a not only provides higher speeds and good range but also decreases wireless LAN deployment costs.

Among the advantages 802.11a has over current wireless LAN technologies are greater scalability, better interference immunity, and significantly higher speed (up to 54 Mbit/s and beyond), which simultaneously allows for higher-bandwidth applications and more users. Chapter 4 provides an overview of how the 802.11a specification functions and its corresponding benefits. It then discusses how existing wireless LAN technologies can be migrated to the 802.11a standard. A cost-effective method of achieving this is to employ a centralized wireless LAN management controller where administrators are presented with a unified management interface for all access points, regardless of radio technology. Because the 5-GHz band is not as widely adopted as the 2.4-GHz band, Chapter 5 addresses the current issues and regulatory aspects of the 5-GHz spectrum allocation in the US, Europe, and the rest of the world. It points out that harmonization efforts with the primary users (particularly satellite users) are an important step in mitigating interference in this band. A key observation in this chapter is the common use of interference mitigation methods such as transmit power control and dynamic frequency selection by different regulatory authorities (e.g., FCC, ITU).

Multimedia remains an integral component of all evolving networks. Applications involving multimedia Internet access and messaging are all influenced by the emergence of multimedia. Because multimedia applications are the most general class of applications, they have the most wide-ranging traffic attributes and communication needs. In addition, these applications impose various performance requirements on the network. There has been increased activity within the IEEE 802 standard committees to create enhancements to existing standards that will be capable of supporting multimedia communications. The IEEE 802.11 Wireless LAN committee is no different. Chapter 6 provides a description of how Quality of Service (QoS) and multimedia applications can be accommodated. This chapter first describes the various wireless committees of the 802 standards and provides useful insights on how these committees function. It then moves on to discuss the current state of proposed changes to the relevant IEEE 802.11 standards and provides context for the reasons behind those proposals. The conclusion of understanding these proposals and the context in which they are made is that the IEEE 802.11 standard is well positioned to meet a growing need for QoS and multimedia support in wireless LANs.

Chapters 7, 8, and 9 focus on wireless LAN security. The free-space wireless link is more susceptible to eavesdropping, fraud, and unauthorized transmission than its wired counterpart. Unauthorized people can tap the radio signal from anywhere within range. If someone sets a mobile client within a wireless coverage area to transmit packets endlessly, all other clients are prevented from transmitting, thus bringing the network down. Being an open medium with no precise bounds makes it impractical to apply physical security as in wired networks. Nevertheless, several security mechanisms can be used to prevent unauthorized access of data transmitted over a wireless LAN. Chapter 7 provides an excellent survey of network security issues in the deployment of wireless LANs. It describes the evolution of security enhancements in the 802.11 standard and discusses the various security threats. The chapter concludes with a proposed solution aimed at solving these security threats. The reader will realize that the original Wired Equivalent Privacy (WEP) algorithm adopted by the 802.11 standard has many security flaws and that the Remote Access Dial-In User Service (RADIUS) and the IEEE 802.1x are two security mechanisms that will play important roles in future 802.11 standards. Chapter 8 by Dorothy Stanley from Agere Systems discusses topics of encryption, key management, and end user authentication, beginning with current issues and available solutions. Unlike Chapter 7, this chapter provides a more in-depth coverage of the technical issues associated with 802.11 network security. Among the important topics covered include media access layer encryption enhancements, the Extensible Authentication Protocol (EAP), and the access server authentication procedure based on session (rather than packet based) encryption. Whereas the 802.11 standard deals with network security primarily through encryption at the physical layer and authentication over a single network link, Chapter 9 advocates the use of virtual private network (VPN) technology that increases the span of network security using the IP layer. VPNs provide three levels of security namely, end-to-end (multiple link) user authentication, encryption, and data authentication. Authentication ensures that only authorized users (over a specific device) are able to connect, send, and receive data over the wireless network. Encryption offers additional protection as it ensures that even if transmissions are intercepted, they cannot be decoded without significant time and effort. Data authentication ensures the integrity of data on the wireless network, guaranteeing that all traffic is from authenticated devices only. From this chapter, the reader will also understand the various forms of VPN application, including deployment in enterprise, public, and home networks.

A compelling use of wireless LANs is in overcoming the inherent limitations of wireless wide area networks (WANs). Current 3G wireless WANs provide data rates of up to 2 Mbit/s and operate on expensive radio spectrum, whereas wireless LANs offer data rates of up to 54 Mbit/s and operate on free unlicensed frequency bands. This led to some technologists predicting that, eventually, we are more likely to see dense urban broadband wireless LANs that are linked together into one network rather than widespread use of high-powered wireless WAN handsets cramming many bits into expensive and narrow slices of radio spectrum. However, wireless LANs offer limited Internet roaming capability and global user management features. The next three chapters show how these weaknesses can be overcome by combining high-speed wireless LANs with the large-scale public infrastructure of mobile cellular networks. In Chapter 10, entitled "Wireless LAN for Mobile Operators," Philippe Laine from Alcatel provides a comprehensive overview of how modern cellular networks can be integrated with wireless LAN systems to enable wide-area Internet access for service providers and users. This chapter first describes the evolution of voice-based cellular networks to data-centric, Internet-based networks. Using many illustrative figures, the author then considers various coupling methods that will combine the reach and mobility features of wide-area cellular networks with the high data rate of wireless LANs. The chapter concludes with a discussion on implementation issues including billing, security, and roaming. Chapter 11 describes a more specific public wireless LAN system. The system efficiently combines wireless LAN access with the widely deployed Global System Mobile/General Packet Radio Service (GSM/GPRS) roaming infrastructure. In addition, the architecture exploits GSM authentication, SIM-based user management, and secured billing mechanisms. This gives the cellular operator a major competitive advantage compared with Internet service providers who have neither a large mobile customer base nor a cellular-type roaming service. We note again the critical role assumed by the RADIUS and 802.1x standards in ensuring network security. The combined cellular and wireless LAN systems described in Chapter 11 more or less operate independently, each network type retaining its unique characteristics for ease of deployment and less dependence on evolving standards. Chapter 12 expands the scope of cooperation between these networks by advocating the use of cellular network features in current wireless LAN systems. It studies the parallels between wireless LANs and cellular networks and highlights the limitations of current wireless LAN standards, which only cater for the physical and media access layers. For improved security, manageability, usability, and performance, additional cellular-type network infrastructure must be incorporated into wireless LANs.

Chapter 13 by Mike Sheppard describes Bluetooth, a short-range wireless communication standard that enables personal area networking among a wide variety of electronic devices, ranging from laptops to cell phones, computers to printers, personal digital assistants to wireless headsets, and many other devices and applications. The lower layers of Bluetooth form the basis of the IEEE 802.15.1 standard on wireless personal area networks, which was approved in March 2002. The chapter discusses the evolution of Bluetooth and the essential technologies encompassing the standard and provides insights on the future success of the standard. Closely associated with this chapter is Chapter 14 by Steve Shellhammer, which discusses the various issues related to Bluetooth and 802.11b coexistence. Coexistence is the ability for multiple protocols to operate in the same frequency band without significant degradation to either's operation. Although Bluetooth and 802.11 coexistence has been a hot area of research for the last two years, an impending amendment to the FCC's Part 15 rules can resolve this problem somewhat. The new rules allow Bluetooth to hop across as few as fifteen 1-MHz channels in the 2.4-GHz band. A frequency-division approach (known as adaptive hopping) essentially allows Bluetooth devices to operate simultaneously with 802.11b devices by segregating the 2.4-GHz frequency band into two nonoverlapping sections, one for Bluetooth and the other for 802.11b. This solution once again demonstrates the FCC's practical mode of operation.

Chapter 15, by Time Domain Corporation, offers a glimpse of a promising wireless technology called ultra wide band (UWB). UWB operates by transmitting pulses over several GHz of bandwidth at very low power (microwatts). Each pulse sequence is pseudorandomly modulated, thus appearing as "white noise" similar to spread spectrum communications. UWB exhibits excellent spectral efficiency because it takes advantage of underutilized spectrum, effectively creating "new" spectrum for existing and future services. The best applications for UWB are indoor use in high-clutter environments. As such, the technology can overcome some of the spectrum and power limitations associated with current wireless LAN systems. UWB technology enables not only communications devices but also positioning/geo-location capabilities of exceptional performance. With the FCC removing the barriers to UWB commercialization in February 2002, the technology will certainly attract much interest in the months ahead.

I am humbled by the immense learning process while editing this book and am indebted to the authors for their diligence in producing some truly fine chapters. Putting these chapters together is like combining the best players from different football teams to form a dream team. I hope you will find ideas expressed in the various chapters thought-provoking and that you will use this book as an impetus for more ground-breaking developments in wireless LAN technologies. Feel free to send your comments to bennybing@ieee.org.

Benny Bing
*Atlanta, Georgia, USA*
*March 2002*

# Acknowledgments

Undertaking this edited book project took more time than I had anticipated. It would have taken much more time if not for the cooperation of the following people. I would like to thank Ronnie Holland, Joia Turner and Mark Terwilliger who have helped produce their company's white paper under short notice. I am grateful to Chris Heegard for his patience in working on the changes to his chapter, to Jouni Mikkonen, James Chen, and Mike Sheppard for graciously expanding the contents of their chapters, and to Dorothy Stanley and Steve Shellhammer for their diligence in finalizing their chapters in a timely manner even while working under extremely tight deadlines. Finally, I am grateful to Phil Meyler, Senior Editor of John Wiley, for his support and encouragement as well as to Kirsten Rohstedt, Susan Adams and Danielle Lacourciere for their help and diligence.

# Chapter 1

# Guide to Wireless LAN Analysis

*WildPackets, Inc.*

The world is going wireless. The prohibitive cost of building wired network infrastructures has paved the way for wireless networking on a global scale. Developing countries, with more sophisticated network and Internet access than ever before, have surged ahead in the utilization of wireless networks so that even the most remote parts of the globe have coverage undreamed of only a few years ago. Although deployment of wireless LANs in the United States has lagged behind the rest of the world, the domestic market is now quickly coming up to speed. It is therefore of critical importance for the corporate network manager to understand not only how the wireless revolution is taking place but how this technological paradigm shift will affect the day-to-day monitoring and management of network data. This chapter provides a brief overview of wireless networks and the 802.11b standard in particular, followed by a discussion of troubleshooting and network maintenance problems and the types of monitoring and analysis required to resolve them.

## 1.1 Introduction to Wireless Networking

The market for wireless communications has grown rapidly since the introduction of 802.11b wireless local area networking standards, which offer performance more nearly comparable to that of Ethernet. Business organizations value the simplicity and scalability of wireless LANs and the relative ease of integrating wireless access and the ability to roam with their existing network resources such as servers, printers, and Internet connections. Wireless LANs support user demand for seamless connectivity, flexibility, and mobility.

As their price/performance and reliability continue to improve, wireless LANs are poised to become a common part of most business networks. In some cases, wireless LAN technology may make up the entire network. Cahners In-Stat Group predicts that the Enterprise segment of the wireless LAN market will grow from $771 million in 1999 to nearly $2.2 billion in 2004 (Wireless LAN Market Analysis, January 2000).

### 1.1.1 The Increasing Need for Wireless LAN Analysis

Although the obvious benefits of wireless LANs gain new converts, the network administration requirements for this relatively unfamiliar technology may not be as widely understood. Sometimes referred to as "wireless Ethernet," the IEEE 802.11 standards are in fact a completely distinct set of technologies with their own peculiar strengths and weaknesses. Maintaining the security, reliability, and overall performance of a wireless LAN requires the same kind of ability to look "under the hood" as the maintenance of a more familiar wired network. In addition, wireless LANs are very often integrated with new or existing Ethernet networks. Naturally, diagnostic and troubleshooting requirements for TCP/IP, IPX, and other higher-level protocols and services do not stop at the end of the wire. Wireless networking presents some unique challenges for the network administrator and requires some new approaches to familiar problems. To see what these are—and why they are— we need to know something about how wireless LANs work.

### 1.1.2 Development of the IEEE 802.11b Standard

In 1997, IEEE approved 802.11, the first internationally sanctioned wireless LAN standard. This first standard proposed any of three (mutually incompatible) implementations for the physical layer: infrared (IR) pulse position modulation, radio frequency (RF) signaling in the 2.4-GHz band using frequency hopping spread spectrum (FHSS), or direct sequence spread spectrum (DSSS). The IR method was never commercially implemented. The RF versions suffered from low transmission speeds (2 Mbit/s). In an effort to increase throughput, IEEE established two task groups to explore alternate implementations of 802.11.

Task Group A explored the 5-GHz band, hoping to achieve throughputs in the range of 54 Mbit/s. The challenges, both to produce low-cost equipment operating at such high frequencies and to reconcile competing international uses of this spectrum, may keep their 802.11a standard from reaching wide distribution before 2002 or 2003 despite its promise of high performance. Task Group B explored more sophisticated spectrum spreading technologies in the original 2.4-GHz band. Their 802.11b standard, published in September 1999, can deliver raw data rates up to 11 Mbit/s. The majority of wireless LAN systems in the market today follow the 802.11b standard.

The 802.11b wireless LAN standard specifies the lowest layer of the OSI network model (physical) and a part of the next higher layer (data link). In addition, the standard specifies the use of the 802.2 protocol for the logical link control (LLC) portion of the data link layer. In this same OSI conceptual model of network stack functionality (Fig. 1.1), protocols such as TCP/IP, IPX, NetBEUI, and AppleTalk exist at still higher layers and utilize the services of the layers underneath.

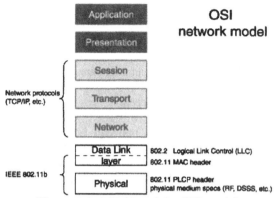

**Figure 1.1.** 802.11 and the OSI model.

### 1.1.3 Radio Frequencies and Channels

The most striking differences between wireless LANs and the more familiar wired networks such as Ethernet are those imposed by the difference in the transmission medium. Whereas Ethernet sends electrical signals through wires, wireless LANs send radio frequency (RF) energy through the air. Wireless devices are equipped with a special network interface card (NIC) with one or more antennae, a radio transceiver, and circuitry to convert between the analog radio signals and the digital pulses used by computers.

Radio waves broadcast on a given frequency can be picked up by any receiver within range tuned to that same frequency. Effective or usable range depends on signal power, distance, and interference from intervening objects or other signals. Information is carried by modulating the radio waves. In spread spectrum technologies, additional information is packed into a relatively small range of frequencies (a section of bandwidth called a channel) by having both sender and receiver use a predetermined set of codes, such that each small modulation of the radio wave carries the greatest possible information. The term "Direct Sequence Spread Spectrum" in DSSS refers to one particular approach to packing more data into a given piece of RF spectrum—more data in the channel.

The FCC in the United States, and other bodies internationally, control the use of RF spectrum and limit the output power of devices. The 802.11b wireless LAN standard attempts to deliver maximum performance within the limits set by these bodies, current radio technology, and the laws of physics. Low output power, for example, limits 802.11b wireless LAN transmissions to fairly short effective ranges, measured in hundreds of yards. In addition, the nature of radio waves and of spectrum spreading technology means that signal quality, and hence network throughput, diminishes with distance and interference. The higher data rates rely on more complex spectrum spreading techniques. These in turn require an ability to distinguish very subtle modulations in the RF signal. To overcome signal degradation problems, 802.11b wireless LANs can gracefully step down to a slower but more robust transmission method when conditions are poor and then step back up again when conditions improve. The full set of data rates in 802.11b wireless LANs is 11, 5.5, 2, and 1 Mbit/s.

The 2.4-GHz band (2.40–2.45 GHz) in US implementations is divided into 11 usable channels. To limit interference, any particular 802.11b wireless LAN network will use less than half of these in operation. All network hardware is built to be able to listen or transmit on any one of these channels, but both sender and receiver must be on the same channel to communicate directly.

### 1.1.4 Collision Avoidance and Media Access

One of the most significant differences between Ethernet and 802.11b wireless LANs is the way in which they control access to the medium, determining who may talk and when. Ethernet uses carrier sense multiple access with collision detection (CSMA/CD). This is possible because an Ethernet device can send and listen to the wire at the same time, detecting the pattern that shows a collision is taking place. When a radio attempts to transmit and listen on the same channel at the same time, its own transmission drowns out all other signals. Collision detection is impossible.

The carrier sense capabilities of Ethernet and wireless LANs are also different. On an Ethernet segment, all stations are within range of one another at all times, by definition. When the medium seems clear, it is clear. Only a simultaneous start of transmissions results in a collision. As shown in Figure 1.2, nodes on a wireless LAN cannot always tell by listening alone whether or not the medium is in fact clear. In a wireless network a device can be in range of two others, neither of which can hear the other but both of which can hear the first device. The access point in Figure 1.2 can hear both node A and node B, but neither A nor B can hear each other. This creates a situation in which the access point can be receiving a transmission from node B without node A sensing that node B is transmitting. Node A, sensing no activity on the channel, may then begin transmitting, jamming the access point's reception of node B's transmission already under way. This is known as the "hidden node" problem.

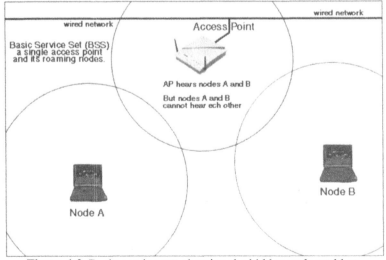

**Figure 1.2.** Basic service set, showing the hidden node problem.

To solve the hidden node problem and overcome the impossibility of collision detection, 802.11b wireless LANs use CSMA/CA (carrier sense multiple access with collision avoidance). Under CSMA/CA, devices use a four-way handshake (Fig. 1.3) to gain access to the airwaves to ensure collision avoidance. To send a direct transmission to another node, the source node puts a short Request to Send (RTS) packet on the air, addressed to the intended destination. If that destination hears the transmission and is able to receive, it replies with a short Clear to Send (CTS) packet. The initiating node then sends the data, and the recipient acknowledges all transmitted packets by returning a short Acknowledgement (ACK) packet for every transmitted packet received.

Timing is critical to mediating access to the airwaves in wireless LANs. To ensure synchronization, access points or their functional equivalents periodically send beacons and timing information.

### 1.1.5 Wireless LAN Topologies

Wireless LANs behave slightly differently depending on their topology or makeup of member nodes. The simplest arrangement is an ad hoc group of independent wireless nodes communicating on a peer-to-peer basis, as shown in Fig. 1.4. The standard refers to this topology as an Independent Basic Service Set (IBSS) and provides for some measure of coordination by electing one node from the group to act as the proxy for the missing access point or base station found in more complex topologies. Ad hoc networks allow for flexible and cost-effective arrangements in a variety of work environments, including hard-to-wire locations and temporary setups such as a group of laptops in a conference room.

The more complex topologies, referred to as infrastructure topologies, include at least one access point or base station. Access points provide synchronization and coordination, forwarding of broadcast packets, and, perhaps most significantly, a bridge to the wired network. The standard refers to a topology with a single access point as a Basic Service Set (BSS), as shown in Figure 1.2. A single access point can manage and bridge wireless communications for all the devices within range and operating on the same channel.

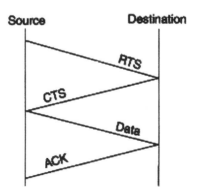

**Figure 1.3.** A four-way handshake ensures collision avoidance in 802.11b networks.

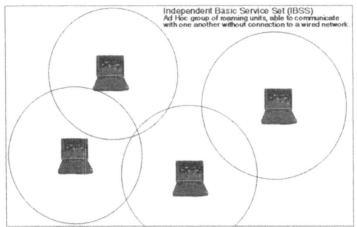

**Figure 1.4.** An IBSS or "ad hoc" network.

To cover a larger area, multiple access points are deployed. This arrangement (shown in Fig. 1.5) is called an Extended Service Set (ESS). It is defined as two or more BSSs connecting to the same wired network. Each access point is assigned a different channel wherever possible to minimize interference. If a channel must be reused, it is best to assign the reused channel to the access points that are least likely to interfere with one another.

When users roam between cells or BSSs, their mobile device will find and attempt to connect with the access point with the clearest signal and the least amount of network traffic. In this way, a roaming unit can transition seamlessly from one access point in the system to another, without losing network connectivity.

An ESS introduces the possibility of forwarding traffic from one radio cell (the range covered by a single access point) to another over the wired network. This combination of access points and the wired network connecting them is referred to as the Distribution System (DS). Messages sent from a wireless device in one BSS to a device in a different BSS by way of the wired network are said to be sent by way of the DS.

*Note*: To meet the needs of mobile radio communications, the 802.11b wireless LAN standard must be tolerant of connections being dropped and reestablished. The standard attempts to ensure minimum disruption to data delivery and provides some features for caching and forwarding messages between BSSs. Particular implementations of some higher-layer protocols such as TCP/IP may be less tolerant. For example, in a network in which DHCP is used to assign IP addresses, a roaming node may lose its connection when it moves across cell boundaries and have to reestablish it when it enters the next BSS or cell. Software solutions are available to address this particular problem. In addition, IEEE may revise the standard in ways that mitigate this problem in future versions.

Whether they have one base station or many, most corporate wireless LANs will operate in infrastructure mode to access servers, printers, Internet connections, and other resources already established on wired networks. Even users seeking an "all wireless" solution may find that an access point does a better job of mediating communications with an Internet connection, for example, and is worth the additional expense.

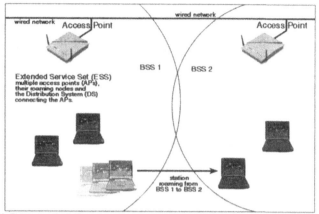

**Figure 1.5.** Extended Service Set supports roaming from one cell to another.

### 1.1.6 Authentication and Privacy

Authentication restricts the ability to send and receive on the network. Privacy ensures that eavesdroppers cannot read network traffic. Authentication can be open or based on knowledge of a shared key. In either case, authentication is the first step for a device attempting to connect to an 802.11b wireless LAN. The function is handled by an exchange of management packets. If authentication is open, then any standards-compliant device will be authenticated. If authentication is based on a shared key, then a device must prove that it knows this shared key to be authenticated.

Wired Equivalent Privacy (WEP) is a data encryption technique supported as an option in the 802.11b wireless LAN standard. The technique uses shared keys and a pseudorandom number (PN) as an initialization vector (IV) to encrypt the data portion of network packets. The 802.11b wireless LAN headers themselves are not encrypted. The designers' purpose in supporting this feature was to give a wireless network, with its inherent vulnerability to eavesdropping, a level of security similar to that enjoyed by a wired network operating without encryption. Eavesdropping on a wired network, they reasoned, requires a physical network tap or a suite of sophisticated radio listening devices. Eavesdropping on a radio network requires only a device capable of listening on the same channel or frequency. Because all 802.11b wireless LAN adapters are capable of listening on any of the usable channels, eavesdropping is almost a certainty, given a large enough number of devices in circulation.

The original WEP specification called for 64-bit key length encryption (often referred to as "40 bit" with respect to the user-defined key). In part, this was an explicit effort to make commercial implementations of the protocol exportable from the US in an era when only the very weakest encryption technologies were granted export licenses. The standard's support for such a weak encryption method also underlines the design function of encryption in this protocol, however. It is intended to stop casual eavesdropping, not to stop a concerted attack. Several vendors now support 128-bit key lengths. This significantly increases the barriers to attack, but even at 128-bit key lengths, WEP is still the door to an office, not a bank vault. Any of these levels of encryption serves the primary purpose of WEP quite well.

Because WEP encrypts all the data above the 802.11b wireless LAN layers, it can prevent network analysis tools from decoding higher-level network protocols and so prevent accurate troubleshooting of problems with TCP/IP, IPX, NetBEUI, and so forth. To overcome this limitation, network analysis tools should allow users to specify the WEP shared key set for their network so they can decode the network data contained in 802.11b wireless LAN packets in the same way that every other station on the user's network does.

*Note*: Although it is possible to implement WEP with open authentication, this is strongly discouraged because it leaves the door open for intruders to collect enough information to compromise the security of WEP.

## 1.2 Packet Structure and Packet Types

Like the rest of the 802 family of LAN protocols, 802.11b wireless LAN sends all network traffic in packets. There are three basic types: data packets, network management packets, and control packets. The first subsection describes the basic structure of an 802.11b wireless LAN data packet and the information it provides for network analysis. The second section describes the management and control packets, their functions, and the part they play in network analysis.

### 1.2.1 Packet Structure

All the functionality of the protocol is reflected in the packet headers. RF technology and station mobility impose some complex requirements on 802.11b wireless LANs. This added complexity is reflected in the long Physical Layer Convergence Procedure (PLCP) headers as well as the data-rich MAC header (Fig. 1.6).

Because 802.11b wireless LANs must be able to form and re-form their membership constantly, and because radio transmission conditions themselves can change, coordination becomes a large issue in wireless LANs. Management and control packets are dedicated to these coordination functions. In addition, the headers of ordinary data packets contain a great deal more information about network conditions and topology than, for example, the headers of Ethernet data packets would contain (Fig. 1.7).

A complete breakout of all the fields in the packet headers and the values they may take is beyond the scope of this chapter. Instead, Tables 1.1–1.5 present the types of information that 802.11b wireless LAN data packet headers convey. The tables also show the types of information carried in management and control packets.

**802.11 packet structure**

| OSI Physical (PHY) layer | | OSI Data Link layer | | higher OSI layers | packet trailer | |
|---|---|---|---|---|---|---|
| PLCP preamble | header | MAC Header | LLC (opt) | Network Data | FCS | End Delimiter |

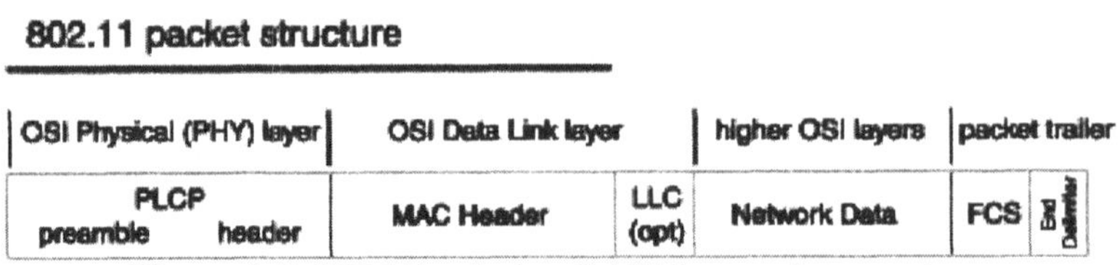

**Figure 1.6.** 802.11b wireless LAN data packet structure.

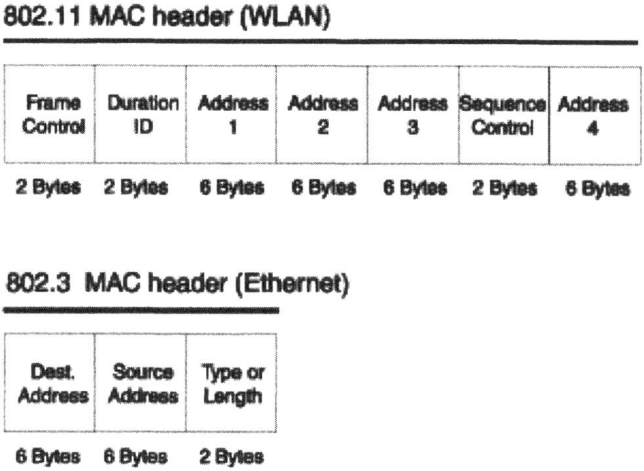

## 802.11 MAC header (WLAN)

| Frame Control | Duration ID | Address 1 | Address 2 | Address 3 | Sequence Control | Address 4 |
|---|---|---|---|---|---|---|
| 2 Bytes | 2 Bytes | 6 Bytes | 6 Bytes | 6 Bytes | 2 Bytes | 6 Bytes |

## 802.3 MAC header (Ethernet)

| Dest. Address | Source Address | Type or Length |
|---|---|---|
| 6 Bytes | 6 Bytes | 2 Bytes |

**Figure 1.7.** Comparison of MAC headers: 802.11b wireless LAN to 802.3 Ethernet.

**Table 1.1. Authentication/privacy protocol functions.**

The first step for a device in joining a BSS or IBSS is authentication. This can be an open or a shared key system. If WEP encryption of packet data is enabled, shared key authentication should be used. Authentication is handled by a request/response exchange of management packets.

| Info Type | Usage |
|---|---|
| Authentication ID | This is the name under which the current station authenticated itself on joining the network. |
| WEP enabled | If this field is true, then the payload of the packet (but not the wireless LAN headers) will be encrypted using WEP. |

**Table 1.2. Network membership/topology protocol functions.**

The second step for a device joining a BSS or IBSS is to associate itself with the group or with the access point. When roaming, a unit also needs to disassociate and reassociate. These functions are handled by an exchange of management packets, but the current status is shown in packet headers.

| Info Type | Usage |
|---|---|
| Association | Packets can show the current association of the sender. Association and Reassociation are handled by request/response management packets. Disassociation is a simple declaration from either an access point or a device. |
| IBSSID or ESSID | The ID of the group or its access point. A device can only be associated with one access point (shown by the ESSID) or IBSS at a time. |
| Probe | These are request/response management packets used by roaming devices in search of a particular BSS or access point. They support a roaming unit's ability to move between cells while remaining connected. |

**Table 1.3. Network conditions/transmission protocol functions.**

| Info Type | Usage |
|---|---|
| The 802.11b wireless LAN protocol supports rapid adjustment to changing conditions, always seeking the best throughput. | |
| Channel | The channel used for this transmission. |
| Data rate | The data rate used to transmit a packet. |
| Fragmentation | 802.11b wireless LANs impose their own fragmentation on packets, completely independent of any fragmentation imposed by higher-level protocols such as TCP/IP. A series of short transmissions is less vulnerable to interference in noisy environments. This fragmentation is dynamically set by the protocol in an effort to reduce the number, or at least the cost, of retransmissions. |
| Synchronization | Several kinds of synchronization are important in wireless LANs. Network management packets called "beacon" packets keep members of a BSS synchronized. In addition, devices report the state of their own internal synchronization. Finally, all transmissions contain a time stamp. |
| Power save | Laptops in particular need to conserve power. To facilitate this, the protocol uses a number of fields in data packets including the power save-poll (PS-Poll) control packet to let devices remain connected to the network while in power save mode. |

**Table 1.4. Transmission control protocol functions.**

| Info Type | Usage |
|---|---|
| Although the protocol as a whole actually controls the transmission of data, certain header fields and control packets have this as their particular job: | |
| RTS, CTS, ACK | These are control packets used in the four-way handshake in support of collision avoidance. |
| Version | The version of the 802.11 protocol used in constructing the packet. |
| Type and subtype | The type of packet (data, management, or control), with a subtype specifying its exact function. |
| Duration | In support of synchronization and orderly access to the airwaves, packets contain a precise value for the time that should be allotted for the remainder of the transaction of which this packet is a part. |
| Length | Packet length. |
| Retransmission | Retransmissions are common. It is important to declare which packets are retransmissions. |
| Sequence | Sequence information in packets helps reduce retransmissions and other potential errors. |
| Order | Some data, such as voice communications, must be handled in strict order at the receiving end. |

**Table 1.5. Routing protocol functions.**

| Info Type | Usage |
|---|---|
| Again, many fields are related to routing traffic, but the following are most directly related. | |
| Addresses | There are four address fields in 802.11b wireless LAN data packets, instead of the two found in Ethernet or IP headers. This is to accommodate the possibility of forwarding to, from, or through the DS. In addition to the normal Destination and Source addresses, these fields may show the Transmitter, the Receiver, or the BSS ID. Which of the address fields shows what address depends on whether (and how) the packet is routed by way of the DS. Control and management packets need only three address fields because they can never be routed both to and from (that is, through) the DS. |
| To/from DS | In an ESS, traffic can be routed from a device using one access point to a device using a different access point somewhere along the wired network. These fields describe routing through the DS and tell the receiving device how to interpret the address fields. |
| More data | Access points can cache data for other devices. This serves both roaming across BSS or cell boundaries and the power save features. When a device receives a message from an access point, it may be told that the access point has more data waiting for it as well. |

### 1.2.2 Management and Control Packets

Control packets are short transmissions that directly mediate or control communications. Control packets include the RTS, CTS, and ACK packets used in the four-way handshake (Fig. 1.3), as well as power save polling packets and short packets to show (or show and acknowledge) the end of contention-free periods within a particular BSS or IBSS. Management packets are used to support authentication, association, and synchronization. Their format is similar to those of data packets, but with fewer fields in the MAC header. In addition, management packets may have data fields of fixed or variable length, as defined by their particular subtype. The types of information included in management and control packets are shown in Tables 1.1–1.5, along with the related information found in data packet headers.

## 1.3 Wireless Packet Analysis

Wireless networks require the same kinds of analytical and diagnostic tools as any other LAN to maintain, optimize, and secure network functions. The 802.11b wireless LAN standard offers even more data to packet analysis than any of the other members of the 802 family of protocols. After a brief note about the kinds of information available to packet analysis, this section describes four broad areas in which protocol analyzers can be of particular use in network troubleshooting and administration.

### 1.3.1 A Note on Packet Analysis and RF Monitoring

Network administrators accustomed to Ethernet may be daunted at first by the unfamiliarity of RF technology. They may wonder if they need RF detection and monitoring devices more than their traditional network analysis tools. It is true that wireless LANs pose unique problems. It is also true that some understanding of, for example, RF signal propagation will be useful–particularly in the initial deployment of larger ESS networks.

That being said, however, the need for specialized hardware to support an 802.11b wireless LAN is no greater than the need for similar equipment in support of the wiring plan of an Ethernet network, and perhaps even less. The reason for this lies in the specification of the standard itself and in the design of the network hardware. Nearly all of the parameters of physical network performance are available directly or indirectly in the packet headers and network management packets themselves. All that is required for most troubleshooting and maintenance tasks is a good network analysis program able to read, collect, and display the data it finds on the network in a clear and meaningful way.

### 1.3.2 Configuration and Traffic Management

One of the advantages of 802.11b wireless LANs is their ability to dynamically adjust to changing conditions and almost to configure themselves to make the best use of available bandwidth. These capabilities work best, however, when the problems they address are kept within limits.

For example, where interference is high, 802.11b wireless LAN nodes will continue to increase fragmentation, simplify spectrum spreading techniques, and decrease transmission rates. Another symptom of high interference is increased retransmissions, especially when they occur despite high fragmentation. Although some network applications may show no ill effects from this condition, others may begin to lag with too many retransmissions of packets already reduced well below their most efficient transmission size. Remember that 802.11b wireless LAN packet headers are quite large. This means high overhead and a low usable data rate when packet fragmentation and retransmissions are both high. If only one or two network applications seem to be affected, it may not be immediately obvious that there is a more general problem. Using a wireless packet analyzer in such a case can quickly determine the state of the network (Fig. 1.8). Possible sources of interference can be examined and the results tested in near real time.

802.11b wireless LAN BSSs and ESSs also have the ability to dynamically configure themselves, associating and reassociating roaming nodes, first with one access point and then with another. The physical location and RF channel used by each access point must be chosen by humans, however. These choices can lead to smooth network functioning or to unexpected problems. To help evaluate network topologies, a packet analyzer must be able to display signal strength and transmission rate for each packet found on a given channel. Furthermore, the user must have control over what channel—better still, which base station—the packet analyzer will scan. With these tools, a packet analyzer can be used to build a picture of conditions at the boundaries between cells in an ESS. Such a survey may find dead spots in a particular configuration or identify places where interference seems to be unusually high. Solving the problem may require changing the channel of one or more access points, or perhaps moving one or more to a new location. The effects of each change can quickly be monitored with a packet analyzer.

### 1.3.3 Identifying Potential Security Problems

Security is a particular concern in wireless networks. Although 128-bit shared key encryption makes WEP secure against casual intrusion, it does not stop eavesdropping. Instead, it seeks to keep eavesdroppers from finding anything of use. The weak link in the security of wireless LANs is not the encryption scheme (particularly not at 128-bit levels) but user authentication. An authenticated user can gain all the information he or she needs to compromise even 128-bit encryption. Given enough computing power, eavesdropping plus the information gathered as an authenticated user will allow any determined attacker to crack WEP.

Packet analyzers cannot detect eavesdroppers. They can detect failed authentication attempts, however. If a packet analyzer has a filtered or triggered start capture system, such an analyzer can be set to scan continuously for failed authentication attempts, capturing all the traffic exchanged in these attempts and making it possible to identify the potential attacker (Fig. 1.9). In a similar way, packet analyzers with sophisticated filtering can be set to watch for WEP encrypted traffic to or from any MAC address that is *not* a known user of the system.

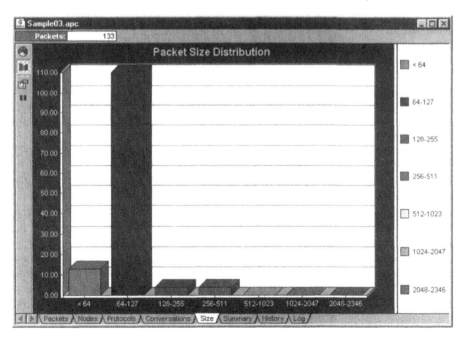

**Figure 1.8.** An unusual predominance of small packets may indicate interference.

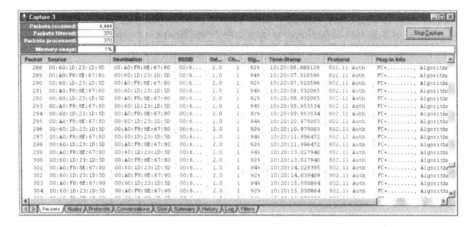

**Figure 1.9.** Focusing on authentication requests to guard against intrusion.

### 1.3.4 Analyzing Higher-Level Network Protocols

Managing a network is more than just managing Ethernet or the wireless LAN. It also means making sure that all the resources users expect to access over the network remain available. This means troubleshooting the network protocols that support these resources. When wireless LANs are used to extend and enhance wired networks, there is no reason to expect that the behavior of higher-level protocols on these mobile clients will be any more or less prone to problems than on their wired equivalents. Although much of this work can be done by capturing traffic from the wired network alone, some problems will yield more quickly to analysis of wireless-originated traffic captured before it enters the DS. To determine whether access points are making errors in their bridging, or whether packets are being malformed at the client source, you must be able to see the packets as they come from the client node.

In an all-wireless environment, the only way to troubleshoot higher-level protocols like IPX and TCP/IP protocols is to capture the packets off the air. In smaller satellite offices in particular, this all-wireless solution is increasingly common. It offers quick setup and can cover areas that would be awkward to serve with wiring, such as noncontiguous office spaces on the same floor. The only wired part of such networks may be the connection from the DSL modem, through the router to the access point.

The actual troubleshooting of these higher-level protocols is no different on a wired or a wireless LAN, provided the network analysis software can read the packets fully. If WEP is enabled, the protocol analyzer must be able to act like any other node on the wireless network and decode the packet payloads using the shared keys. Theoretically, of course, WEP is only an option and can be temporarily disabled. In practice, it is both unlikely and inadvisable that any wireless LAN should operate without WEP. Nor is it particularly simple to turn this function on and off at will. The ability to use WEP in the same way as all other nodes on the network must be built into the analyzer.

### 1.3.5 Roaming and Wireless Analysis

The 802.11b wireless LAN standard leaves much of the detailed functioning of the DS to others. This was a conscious decision on the designers' part, because they were most concerned to make their standard entirely independent of any other existing network standards. As a practical matter, an overwhelming majority of 802.11b wireless LANs using ESS topologies are connected to Ethernet LANs and make heavy use of TCP/IP. These users, at least, would probably have favored a bit more interconnection, even at the cost of some independence.

Wireless LAN vendors have stepped into the gap to offer proprietary methods of facilitating roaming between nodes in an ESS. Third-party software is also available to cache and proxy for roaming nodes at the TCP/IP layer. Although no packet analyzer is likely to recognize all—or perhaps any—of these proprietary approaches out of the box, some packet analyzers can be taught to recognize and decode new packet types. Others can be taught how to capture and filter packets on the basis of particular values or strings found at specific locations within packets. Either of these approaches would deliver the ability to diagnose and troubleshoot a wider variety of performance problems in transitioning nodes from one BSS to another (Fig. 1.10).

## 1.4 Conclusion

The competitive advantage of wireless networks is evident in the flexibility, mobility, and interoperability of wireless LANs standing alone or in conjunction with existing conventional networks. As corporate operations increasingly rely on wireless networks for effective and real-time communication, the ability to analyze and troubleshoot network problems in the wireless environment becomes critical to the management of network resources.

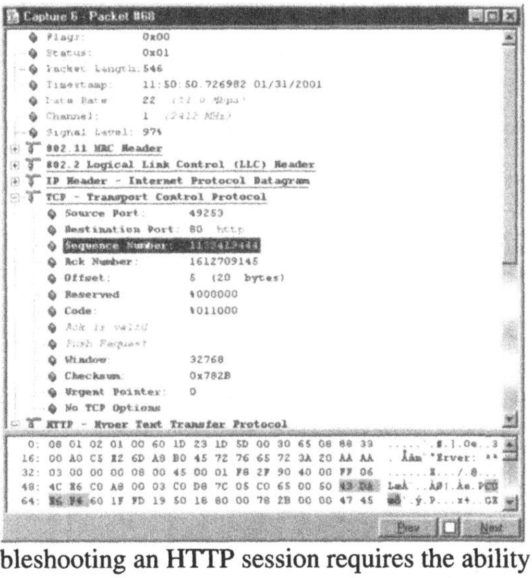

**Figure 1.10.** Troubleshooting an HTTP session requires the ability to read packet data.

## Appendix: AiroPeek Protocol Analyzer

WildPackets' AiroPeek Protocol Analyzer is specifically designed to meet the challenges of network management in the wireless environment. AiroPeek is a comprehensive packet analyzer based on the IEEE 802.11b protocol standard for wireless communications, supporting all standard higher-level network protocols such as TCP/IP, AppleTalk, NetBEUI, and IPX (Fig. 1.11). AiroPeek capabilities include:

- Packet capture: direct, full, filtered, triggered, alarmed, or any combination;
- Shows data rate, channel, and signal strength for each packet;
- Full 802.11b wireless LAN protocol decodes;
- Support for WEP decryption (with user-supplied key sets);
- Continuous monitoring of network statistics in real time;
- Full packet decodes (requires WEP keys, where WEP is enabled);
- Statistical analysis: for all traffic and for specific sets of captured packets;
- Filters, standard and user defined, including use of logical AND, OR, and NOT;
- Name table, caches found names, substitutes user-defined or card vendor names;
- Customized output of packet data, as lists or decodes;
- Alarms, triggers, and notifications, all user definable;
- Customized output of statistics (HTML, XML, text).

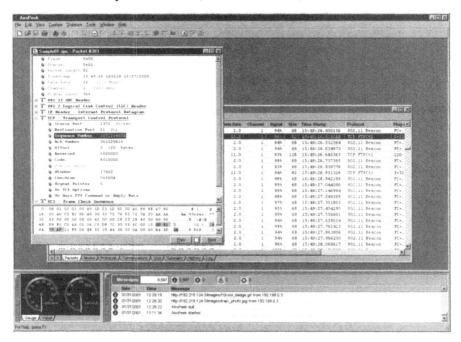

**Figure 1.11.** AiroPeek network analysis tools for 802.11b wireless LANs.

# Chapter 2

# Evolution of 2.4-GHz Wireless LANs

Chris Heegard, Ph.D., John (Seán) T. Coffey, Ph.D., Srikanth Gummadi,
Peter A. Murphy, Ph.D., Ron Provencio, Eric J. Rossin, Ph.D., Sid Schrum,
and Matthew B. Shoemake, Ph.D.
*Texas Instruments, Inc.*

This chapter considers the recently successful IEEE 802.11b standard for high-performance wireless Ethernet and a proposed extension that provides for 22 Mbit/s transmission. IEEE 802.11 sets standards for wireless Ethernet or wireless local area networks. The chapter describes the history of the IEEE 802.11 standards and the market opportunities in the wireless Ethernet field. The chapter gives a brief description of the Medium Access Control (MAC) layer and then presents details about the physical layer methods, including coding descriptions and performance evaluations. The chapter also discusses the role and limitations of spread spectrum communications in wireless Ethernet. A comparison in terms of range versus rate with several alternatives is presented.

## 2.1   Introduction to Wireless Ethernet

In the fall of 1999 a new high-speed standard for wireless Ethernet was ratified by the IEEE 802.11 standards body [1]. This standard extended the original 1 and 2 Mbit/s direct sequence physical layer transmission standard, [2], to break the 10 Mbit/s barrier. The standard, "IEEE 802.11b," established two forms of coding that each deliver both a 5.5 Mbit/s and a 11 Mbit/s data rate.

The second optional choice of coding is known as *packet binary convolutional coding* or "PBCC." This PBCC option was developed by Alantro Communications, now a part of Texas Instruments, Inc. This chapter describes the evolution of the standards for the 2.4-GHz ISM band and the extensions developed by Alantro Communications. The Alantro PBCC system maintains a 22-Mbit/s data rate in the same

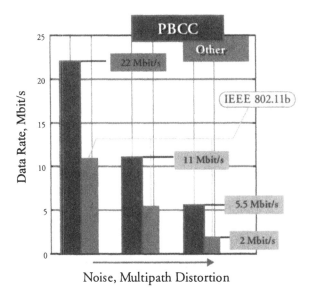

**Figure 2.1.** Performance wireless Ethernet.

environment as the basic 11-Mbit/s system of the current IEEE 802.11b standard as schematically described in Figure 2.1. The system provides for a backward-compatible migration of 802.11b networks into the realm of higher throughput. This chapter greatly expands and augments the material found in an earlier article by Heegard et al. [3].

### 2.1.1    The History and State of the Standards and Marketplace

The origins of wireless networking standardization can be traced to the late 1980s, when members of the IEEE 802.4 standards body considered extensions of token bus technology to wireless transmission. This activity was motivated by FCC spread spectrum regulations that provided for unlicensed transmission in an 83-MHz band of radio frequencies in the 2.4-GHz range. Although this activity did not produce a standard, the interest in these developments led to the creation of IEEE 802.11 in May 1989. The charter for this group is the creation of internationally applicable standards for wireless Ethernet.

The initial standards activity was very contentious, and progress was slow. In addition, as is often the case with good ideas, the technology available for the creation of robust, high-performance/low-cost solutions was not mature. In October of 1997, the first completed standard from the IEEE 802.11 body was ratified. Although the effort to develop the standard was tortuous and time-consuming, the results are impressive. The standard set in 1997 defined both a common *medium access control* (MAC) mechanism as well as three *physical access methods* (PHYs). The three PHYs involved two radio transmission methods for the 2.4-GHz band: *frequency hopping* (FH) and *direct sequence spread spectrum* (DSSS). Both of these PHYs operated as a 1 and 2 Mbit/s data rate and have been deployed in products that were sold on the open market. [The third IEEE 802.11 PHY is an *infrared* (IR) scheme; it is unclear whether any products have been produced with this technology.]

As the first standard was wrapping up, the creation of a new standards activity in IEEE 802.11 was begun. The motivation was to improve the physical layer specification to improve the data rate and throughput parameters of wireless Ethernet. There was strong consensus in the group that wireless Ethernet must be able to deliver a data rate that is comparable to the data rate offered by traditional Ethernet, 10 Mbit/s. It was also agreed that the new activity would concentrate on the physical layer and that changes to the common IEEE 802.11 MAC would be limited to the additions required to make the MAC aware of the parameters of the new PHY technology.

This new activity consisted of two initiatives. The first group considered the definition of a PHY for the unlicensed 5-GHz bands. This effort resulted in the IEEE 802.11a PHY for the 5-GHz band; this standard incorporates a coded multicarrier scheme known as OFDM. The second effort produced a standard commonly known as the IEEE 802.11b standard. This standard offers a DSSS backward-compatible transmission definition that added two new data rates, 5.5 and 11 Mbit/s, as well as two forms of coding. The mandatory coding mode is known as "CCK" modulation and is described in detail in Section 2.3.3.1. The optional code, known as "PBCC" and referred to as the "high-performance mode" of the standard, is described in Section 2.3.3.2. This standard is clearly the most successful standard of the IEEE 802.11 to date; today there are millions of "11b"-compliant devices in the hands of consumers.

Recently, the main standards setting activities of the IEEE 802.11 committee involve enhancements to the MAC, "11e," and even higher rate extensions to the existing standard, "11g." The former activities are directed toward enhancing the MAC, most importantly to improve *quality of service* (QoS) and security. The latter activity was motivated by the work of Alantro Communications, which is a central topic of this chapter (see Sections 2.3.4.1 and 2.4). The main objective of this activity is to define a backward-compatible extension to the existing "11b" networks in a way that improves the data rate (>20 Mbit/s) and overall user experience and satisfaction with wireless Ethernet.

As organizations such as the IEEE 802 Committee continue to push the envelope on the technology front, other organizations are also playing a key role in the adoption of wireless Ethernet technology. The *Wireless Ethernet Compatibility Alliance* (WECA) is the most notable of these organizations. Both the IEEE and WECA have been instrumental in advocating innovation and enhancements to the standard, which has helped fuel rapid industry adoption. WECA's mission is to certify interoperability of 802.11 (IEEE 802.11) products and to promote 802.11b as the global wireless *local area network* (LAN) standard across all market segments. The alliance recently announced that 67 products have passed the rigorous 802.11b certification testing; this makes 802.11b the world's leading wireless LAN standard. Furthermore, momentum continues growing as WECA attracts new members from around the world.

Until 2000 there were several wireless LAN standards competing for the home market; however, 802.11b has resolved this issue. In less than a year, 802.11b has become the single wireless LAN standard for the home, small business, enterprise, and public access areas.

### 2.1.2    Commercial Opportunities

The wide-scale availability of broadband to many homes and most businesses is accelerating the demand for wireless Ethernet. Now that users have easy access to these high-speed communications pipes, they are searching for a simple and cost-effective way to fully utilize them. In homes, a residential wireless gateway can interconnect desktop PCs, telephones, PDAs and other devices with 802.11b-based wireless Ethernet. Soon, entertainment appliances like televisions, stereos, and home theater systems will also be easily connected through this gateway. In the enterprise, users today are able to roam throughout their facilities while maintaining a wireless connection to the organization's network and servers.

As operators continue to roll out broadband services, they face a challenge with many customers. Although bringing high bandwidth to the doorstep is not the hurdle it once was, finding ways to effectively distribute that bandwidth once it crosses the demarcation point poses a mystery to some residential consumers. This broadband access distribution problem impacts small and medium-sized businesses as well. Solving this challenge has the potential to create tremendous market opportunity for communication services companies.

As home networking has gained momentum among consumers, communication companies have faced the challenge of installing new wires in their customers' homes. For example, many older homes have been particularly hard pressed to accommodate traditional Ethernet or LAN wiring; more often than not, the cost of installing it has been prohibitive. Even if it is physically feasible to rewire an existing structure, installing new cabling has meant disruptions and lost productivity in the workplace or at home, in addition to being a major expense.

Recently deployed home networking technology, such as home phone networking, suffers from low user acceptance due to the inconvenience of the technology. It is often the case that the existing phone outlets installed in the home do not match the desired locations for the networked equipment. There are also conflicts in the use of the existing phone wires as the popularity of broadband access methods such as ADSL become more popular.

However, there is a more attractive solution—one that is rapidly gaining acceptance—*wireless Ethernet*. The challenge for communication service companies is to offer the best broadband distribution products to their users. Wireless networking systems are rapidly becoming more and more affordable and the preferred choice for consumers. Recent developments surrounding a proposed performance extension to the wireless Ethernet specification (IEEE 802.11b) hold great promise for an alternative to traditional wired networking. In fact, as the per-user cost of *wireless LANs* is anticipated to drop sharply over the coming years, the market is likewise expected to explode, growing from $624 million in 1999 to $3 billion by 2002, according to Cahners/In-Stat [4].

For communication service companies, all of these performance improvements mean more robust wireless Ethernet installations. High data rates will not only accommodate today's most demanding applications, such as graphically intense interactive gaming or high-definition television, but higher-performance wireless Ethernet installed today will have the performance headroom it needs to accommodate new, even more demanding applications that have yet to be invented. A high-performance wireless Ethernet has the inherent scalability it will need to meet escalating application requirements for years to come.

Advanced technologies have expanded the effective operational range of 802.11b wireless LANs. Users have greater freedom to roam an environment and still be assured that their wireless device will be able to maintain a connection to the network. This can be extremely important for users of all sorts of devices, such as notebook computers, PDAs, or even wireless bar code readers that are used frequently in warehouses or retail locations for inventory management. Highly efficient wireless Ethernet technology promises to make effective use of these broadband pipes, in addition to being an enabler of new and exciting applications. Multimedia applications like high-definition digital streaming video, cordless VoIP telephony, music distribution, connected always-on PDAs, and other appliances are concepts that are just now beginning to tap into the potential that lies beneath the surface of wireless networking technology. Innovation, which has led to the availability of these high-performance, next-generation wireless Ethernet products, is fulfilling the promise of broadband communication for consumers.

## 2.2    Wireless Ethernet Background

The IEEE 802.11 wireless LAN standard, commonly referred to as "wireless Ethernet," is part of a family of IEEE local and metropolitan networking standards, of which 802.3 ("Ethernet") and 802.5 (Token Ring Local Area Network) are two well-known, widely deployed examples (Fig. 2.2). The IEEE 802 standards deal with the Physical and Data Link layers in the ISO *Open Systems Interconnection* (OSI) Basic Reference Model. IEEE 802 specifies the Data Link Layer in two sublayers, *Logical Link Control* (LLC) and *Medium Access Control* (MAC). The IEEE 802 LAN MACs share a common LLC layer (IEEE standard 802.2) and Link Layer address space utilizing 48-bit addresses.

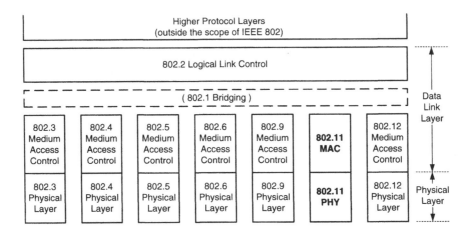

**Figure 2.2.** IEEE 802 Standards.

It is relatively straightforward to bridge between IEEE 802.11 wireless LANs and IEEE 802 wired LANs and to construct extended interconnected wired and wireless 802 LAN networks. Through this means all the services typically offered on wired LANs, such as file sharing, E-mail transfer, and internet browsing, are made available to wireless stations. Transparent untethered LAN connectivity, high data rates (currently 11 Mbit/s and increasing to 22 Mbit/s as described in this chapter), acceptable cost, as well as the inherent interoperability afforded by an international standard are contributing factors to the rapidly increasing popularity of 802.11b wireless LANs.

### 2.2.1   Wireless Ethernet Topology

IEEE 802.11 mobile stations (end user client stations) may be mobile, portable, or stationary. Mobile stations dynamically associate with wireless LAN cells, or *Basic Service Sets* (BSSs). The 802.11 MAC protocol supports the formation of two distinct types of BSSs.

The first type is the independent BSS, or "ad hoc" BSS. Ad hoc BSSs are typically self-forming; they are created and maintained as needed without prior administrative arrangements, often for specific purposes (such as transferring a file from one personal computer to another). Stations in an ad hoc BSS establish MAC layer wireless links with those stations in the BSS with which they desire to communicate, and frames are transferred directly from source to destination stations. Therefore, stations in an ad hoc BSS must be within range of one another to communicate. Furthermore, no architectural provisions are made for connecting the ad hoc BSS to external networks, so communications is limited to the stations within the ad hoc BSS.

The second type of BSS is the infrastructure BSS; this is the more common type used in practice. This type supports extended interconnected wireless and wired networking. Within each infrastructure BSS is an *Access Point* (AP), a special central traffic relay station that normally operates on a fixed channel and is stationary. APs connect the infrastructure BSS to an IEEE 802.11 abstraction known as the *Distribution System* (DS). Multiple APs connected to a common DS form an *Extended Service Set* (ESS). The IEEE 802.11 standard portal function connects the DS to non-802.11 LANs, and ultimately to the rest of the network system, if present. The DS is responsible for forwarding frames within the ESS, between APs and portals, and it may be implemented with wired or wireless links (Fig. 2.3).

The ESS allows wireless LAN connectivity to be offered over an extended area, such as a large campus environment. APs may be placed such that the BSSs they service overlap slightly to provide continuous coverage to mobile stations. In practice DSs are typically implemented by using ordinary wired Ethernet. Commercially available APs include an embedded Ethernet portal, and they are therefore essentially wireless LAN-to-Ethernet bridges.

Mobile stations in an infrastructure BSS establish MAC layer links with an AP. Furthermore, they only communicate directly to and from the selected AP. The AP/DS utilizes store-and-forward retransmission for intra-BSS traffic to provide connectivity between the mobile stations in a BSS. Typically, at most a small fraction of the frames flow between mobile stations within an infrastructure BSS; therefore, retransmission results in a small overall bandwidth penalty. The effective

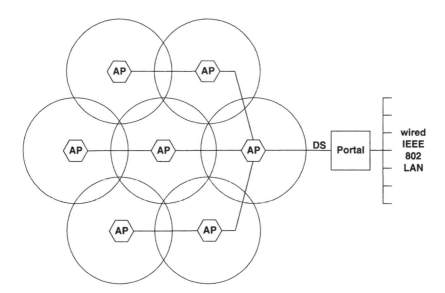

**Figure 2.3.** IEEE 802.11 Network.

physical span of the BSS is on the order of twice the maximum mobile station-to-station range; mobile stations must be within range of the AP to join a BSS but may not be within range of all other mobile stations in the BSS.

Mobile stations utilize the 802.11 architected scan, authentication, and association processes to join an infrastructure BSS and connect to the wireless LAN system. Scanning allows mobile stations to discover existing BSSs that are within range. APs periodically transmit beacon frames that, among other things, may be used by mobile stations to discover BSSs. Before joining a BSS, a mobile station must demonstrate through authentication that it has the credentials to do so. The actual BSS join occurs through association. Mobile stations may authenticate with multiple APs but may be associated with only one AP at a time. Roaming mobile stations initiate handoff from one BSS to another through reassociation. The reassociation management frame is both a request by the sending mobile station to disassociate from the currently associated BSS and a request to join a new BSS.

### 2.2.2  Medium Access Control

The IEEE 802.11 MAC is similar to wired Ethernet in that both utilize a "listen before talk" mechanism to control access to a shared medium. However, the wireless medium presents some unique challenges not present in wired LANs that must be dealt with by the 802.11 MAC. The wireless medium is subject to interference and is inherently less reliable. The medium is susceptible to possible unwanted interception. Wireless networks suffer from the "hidden station" problem: a station transmitting to a receiving station may be interfered with by a third "hidden" station that is within range of the receiver but out of range of the transmitter and therefore does not defer. Finally, wireless stations cannot reliably monitor the idle/busy state of the medium while transmitting. The 802.11 MAC protocol is designed to provide robust, secure communications over the wireless medium.

The fundamental access mechanism is *Carrier Sense Multiple Access/Collision Avoidance* (CSMA/CA) with truncated binary exponential backoff. A station with a frame to transmit contends for the medium by first sensing the medium and deferring until it is idle for a minimum period of time, at which point the station transmits a frame. If the frame is a unicast frame and is received without error by the destination station, the destination station immediately returns a positive acknowledgment frame. If the originating station does not successfully receive the acknowledgment frame, the station assumes that a collision or other event producing a lost packet has occurred. In response to a lost packet, the transmitting station selects a random backoff interval from a uniform range. The range is doubled for every lost packet experienced, until an administratively configurable maximum is reached. The transmitting station then requeues the frame for transmission and contends for the medium after the backoff interval has been satisfied. Multicast and broadcast frames do not use the acknowledgment protocol, and other mechanisms provide protection from lost packets for these frames.

Multiple MAC layer mechanisms contribute to collision avoidance and efficient use of the wireless medium. In contrast to wired Ethernet, if the medium is sensed busy for the first transmission attempt a random backoff is selected and applied. In addition, the backoff counters in deferring 802.11 stations are not decremented when the medium is sensed busy. These two mechanisms reduce the probability of contention when it is most likely to occur, immediately after a transmission.

In addition to the basic contention access mechanism described above, IEEE 802.11 offers an optional contention-free access method. Contention-free access is available only in infrastructure BSSs. Currently, contention-free access is not commonly utilized. However, contention-free access is expected to play an increasingly important role in the future for implementation of quality of service.

With contention-free access, APs gain and maintain control of the wireless medium for extended periods using virtual carrier sense and IFS timing hierarchy (described below). During contention-free periods, polling by the AP is used to grant to mobile stations access to the medium.

The IEEE 802.11 MAC adheres to a strict *interframe space* (IFS) timing hierarchy; four different IFS durations are specified, separated by a minimum of one slot time. These IFS durations establish the length of the gap between nondeferred transmissions, both for frame burst from a single station and for listen-then-talk transmissions. Because of the listen-then-talk access method, transmissions utilizing a given IFS preempt, without contention, those queued transmissions using a longer IFS.

Two types of IFSs, the SIFS and the PIFS, are applied when normally only one station in the BSS has permission to transmit and are therefore intended to result in contention-free access. The *short interframe spacing* (SIFS) is the smallest IFS and it is used betweeen certain multiframe exchange sequences, such as acknowledgement frames sent in response to the error-free reception of a unicast frame. The remaining IFS intervals in order of increasing duration are the DIFS, used by APs to gain priority access to the medium, the PIFS, used by contending stations whose backoff interval has been satisfied, and the EIFS, an IFS enforced after an erroneous reception.

*Virtual carrier sense* is a MAC layer mechanism that augments the physical carrier sense generated by the PHY layer. The duration/ID field in the MAC frame header indicates the expected time remaining to complete the current frame exchange sequence. Stations defer based on previously received duration values, even if the physical carrier sense indicates that the medium is idle. Virtual carrier sense mitigates the hidden station problem. For example, virtual carrier sense prevents a station that is within range of a transmitting station, but out of range of the destination station, from colliding with the acknowledgement frame returned by the destination station.

Virtual carrier sense together with the *Request to Send/Clear to Send* (RTS/CTS) protocol allows stations to place a reservation on the medium before transmitting a data frame. Because RTS and CTS are short control frames and therefore occupy the medium for a relatively short time, the RTS/CTS protocol increases the probability of successful transmission and reduces loss of network throughput due to collisions.

Fragmentation and Automatic Rate Fallback improve the robustness of 802.11 networks by increasing the probability of successful packet transfer, especially for networks that experience time-varying external interference. Long transmit packets presented to the 802.11 MAC by upper protocol layers may be fragmented by the MAC into smaller packets for transmission on the medium. The smaller fragmented packets occupy the medium for shorter periods, increasing the probability that they will be successfully received. Receiving stations reassemble fragments to regenerate the original transmit packet for the upper protocol layers. 802.11 MAC fragmentation is transparent to upper protocol layers.

The 802.11 wireless PHYs provide for multiple transmission bit rates. Generally speaking, lower bit rates enjoy greater range and decreased susceptibility to interference. With automatic rate fallback, 802.11 MACs automatically dynamically choose target station-specific transmission rates based on packet loss statistics and receive signal quality indications provided by the PHY to optimize throughput.

### 2.2.3  Security

Wireless LANs are subject to possible unwanted monitoring. For this reason, IEEE 802.11 specifies an optional MAC layer security system known as *Wired Equivalent Privacy* (WEP). As the name implies, WEP is intended to provide to the wireless Ethernet a level of privacy similar to that enjoyed by wired Ethernets. WEP involves a shared key authentication service with RC4 encryption. The RC4[1] is stream cipher is used to generate a pseudo-random sequence that is "XOR-ed" into the data stream (*ala* a "one time pad"). A key, derived by combining a secret key and an *initialization vector* (IV), is used to set the initial condition or state of the RC4 pseudo-random number generator. By default, each BSS supports up to four 40-bit keys that are shared by all the stations in the BSS. Keys unique to a pair of communicating stations and direction of transmission may also be used (that is, unique to a transmit/receive address pair). Key distribution is outside the scope of the standard but presumably utilizes a secure mechanism.

When a station attempts to authenticate with a second station that implements WEP, the authenticating station presents challenge text to the requesting station. The requesting station encrypts the challenge text using the RC4 algorithm and returns the encrypted text to the authenticating station. The encrypted challenge text is decrypted and checked by the authenticating station before completing authenti-

[1]RC4 is a stream cipher designed by Ronald Rivest for RSA Data Security (now RSA Security). It is commonly known as "Ron's cipher 4."

cation. After authentication (and association), the Frame Body (the MAC payload) is encrypted in all frames exchanged between the stations. Encrypted frames are decrypted and checked by the MAC layer of receiving stations before being passed to the upper protocol layers.

### 2.2.4    The 802.11 MAC Frame Format

The general 802.11 MAC frame format is shown in Figure 2.4. (Not shown is the PHY header that is appended to the front of every frame transmission; see Fig. 2.5.) The Address 2, 3, and 4, Sequence Control, and Frame Body fields are not found in every frame. The frame control field is 16 bits in length, and it contains basic frame control information, including the frame type (data, MAC control, or MAC management) and subtype, if the frame is originated from or is bound to the DS and if the frame is encrypted. The duration/ID field normally indicates the duration of the remainder of a frame exchange sequence and is used to control the virtual carrier sense mechanism as previously described.

The address fields, if present, contain one of the following 48-bit IEEE 802 Link Layer addresses: Destination Address, Source Address, Receiver Address, Transmitter Address, *Basic Service Set ID* (BSSID). For infrastructure networks, the BSSID is the Link Layer address of the AP. For ad hoc networks, the BSSID is a random number generated at the time the ad hoc network is formed. The Receiver, Transmitter, and BSSID addresses are the MAC addresses of stations joined to the BSS that are transmitting or receiving the frame over the wireless Ethernet. Destination and Source addresses are the MAC addresses of stations, wireless or otherwise, that are the ultimate destination and source of the frame. In those cases where two addresses are the same (for example, the Receiver station and the Destination station are one and the same), then a single address field is used. Four address fields are present only in the uncommon case where the DS is implemented with an 802.11 wireless Ethernet, and only for frames traversing the DS. A more typical case involves a frame originating from a wireless station in an infrastructure BSS that is bound for a station on a wired network such as an IEEE 802.3 wired Ethernet. In this situation, the Address 1 field contains the BSSID, the Address 2 field contains the address of the source/transmitter station, the Address 3 field contains the address of the destination station, and the Address 4 field is not present. Includ-

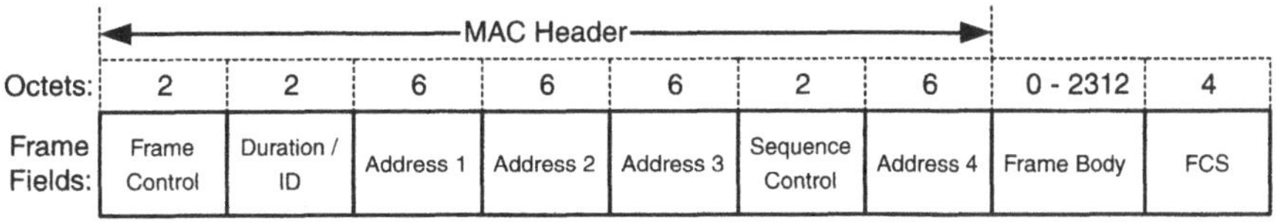

| | | | | | MAC Header | | | | | |
|---|---|---|---|---|---|---|---|---|---|---|
| Octets: | 2 | 2 | 6 | 6 | 6 | 2 | 6 | 0 - 2312 | 4 |
| Frame Fields: | Frame Control | Duration / ID | Address 1 | Address 2 | Address 3 | Sequence Control | Address 4 | Frame Body | FCS |

**Figure 2.4.** Wireless Ethernet frame.

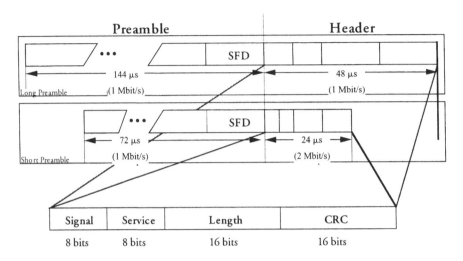

**Figure 2.5.** The physical layer preamble.

ing both the BSSID and the Destination Address (or Source Address for frames flowing to the BSS) in the frame avoids requiring the AP to maintain a list of MAC addresses of stations that are not in the BSS.

The Sequence Control field is 16 bits in length, and it contains the Sequence Number and Fragment Number subfields. Receiving stations use this field to properly reassemble multifragment frames and to identify and discard duplicate frame fragments.

The Frame Body is an optional field that contains the MAC frame payload. For 802.11 MAC management type frames, the Frame Body contains information elements that are specific to the subtype. The FCS field contains a 32-bit *cyclic redundancy check* (CRC). The CRC calculation includes all the MAC frame fields.

## 2.3    The Physical Layer: Coding and Modulation

### 2.3.1    The Physical Layer Preamble

The IEEE 802.11b standard defines a physical layer (PHY) preamble that is transmitted before the wireless Ethernet frame depicted in Figure 2.4. The PHY preamble, as shown in Figure 2.5, consists of a preamble and a header. The header consists of three fields, the *Signal* field, the *Service* field, and the *Length* field. These three fields are protected with a 16-bit CRC that is used to detect transmission errors in the header.

The PHY preamble provides for—

- Packet Detection and Training:
  *The preamble is used to detect the presence of a packet transmission, to decide on antenna selection, and to estimate packet parameters such as signal level for automatic gain control (AGC), carrier offset for frequency tracking, symbol timing, etc.*
- Detection of Frame Boundary (SFD):
  *For packet frame synchronization.*
- Description of Packet Body Modulation and Coding:
  *The choice of coding and modulation is described by the* Signal *field.*
- Virtual Carrier Sense:
  *The* Length *field describes the length of transmission for the body of the packet. This* Length *field measures the transmission in time duration (rather than bits); it is used to initialize a timer in each receiver that detects the packet and is used to time the transmission period. This allows unintended receivers, which may be incapable of demodulating/decoding the type of packet specified in the* Signal *field, to refrain from transmission during the duration of the packet. This mechanism avoids packet collisions and allows for the introduction of new forms of coding and modulation, in an existing network, in a backward-compatible way.*

The original DSSS (1 and 2 Mbit/s) standard defined a PHY preamble with a length of 192 $\mu$s; this preamble is encoded using the 1 Mbit/s encoding method described in Section 2.3.2.1. The "11b" standard added an optional "short preamble" with a duration that is half as long, 96 $\mu$s. The short preamble uses a shorter, 1 Mbit/s encoded preamble, followed by a 2 Mbit/s encoded header.

### 2.3.2  The Low-Rate DS Standards: *The Past*

The original low-rate *direct sequence* (DS) modulation forms a basis for the high-rate extension. This method of coding and modulation is used for preamble generation in all rates and coding combinations. The low-rate system is a direct sequence spread spectrum signal with a "chip rate" of 11 MHz and a data rate of 1 Mbit/s (BPSK) or 2 Mbit/s (QPSK).

*2.3.2.1  Barker 1 and 2 Mbit/s*    The basis for the original 1 and 2 Mbit/s transmission is the incorporation of an 11-bit Barker code (or sequence)

$$B_{11} = [-1, +1, -1, -1, +1, -1, -1, -1, +1, +1, +1]$$

with QPSK or BPSK modulation. This code has the desirable property that the autocorrelation function is minimal (0 or $-1$) except at the origin (where it is 11) as seen in Figure 2.14 in Section 2.4.1. This means that the modulated waveform essentially occupies the same spectrum (see Section 2.4.1) as an 11-MHz uncoded chip signal and that a matched filter receiver, matched to the Barker sequence, will experience a processing gain of $11 = 10.41$ dB.

From a coding point of view, the Barker code can be described in terms of a *linear block code* over the set of integers modulo 4, $Z_4 = \{0, 1, 2, 3\}$. Consider the $k \times n = 1 \times 11$ repetition *generator matrix*

$$G = [1, 1, 1, 1, 1, 1, 1, 1, 1, 1, 1]$$

and the length 11 *cover vector*

$$\mathbf{b} = [2, 0, 2, 2, 0, 2, 2, 2, 0, 0, 0]$$

Then the four Barker codewords for the 2 Mbit/s case are generated by the codeword equation

$$\mathbf{c} = m \cdot G + \mathbf{b} = [c_1, c_2, c_3, c_4, c_5, c_6, c_7, c_8, c_9, c_{10}, c_{11}] \quad \text{(modulo 4)} \quad (2.1)$$

where the message symbol $m \in Z_4$. The transmitted signal is generated with the QPSK mapping in Table 2.1, which produces the signal vector

$$\mathbf{c} \rightarrow \mathbf{x} = [x_1, x_2, x_3, x_4, x_5, x_6, x_7, x_8, x_9, x_{10}, x_{11}] \quad (2.2)$$

The complex values in Table 2.1 are used to represent the "in phase" (the real part or "cosine") and "quadrature phase" (the imaginary part or "sine") of the pulse amplitude modulated carrier. For example, the $A_k$ values in Eq. (2.6), on page 45, would take on the four complex QPSK values given in Table 2.1. The resulting complex baseband signal $x(t)$ would be modulated to a prescribed carrier frequency in the 2.4- to 2.483-GHz range. The transmitter would "mix" (i.e., multiply) the real part of $x(t)$ by a cosine wave at the carrier frequency, mix the imaginary part of $x(t)$ by a sine wave (the cosine wave shifted by 90°), and add the two mixed signals together. This composite signal would be a band-pass signal centered at the carrier frequency as described by Proakis [5].

**Table 2.1.  QPSK Mapping (CCK).**

| Code Symbol $c_i$ | Signal $x_i$ |
|:---:|:---:|
| 0 | $+1 + i$ |
| 1 | $-1 + i$ |
| 2 | $-1 - i$ |
| 3 | $+1 - i$ |

Note that the 2 Mbit/s Barker code is 90° rotationally invariant (i.e., the rotation of a codeword vector **x** by 90° is another codeword). This follows from the fact the addition of 1 (modulo 4) to a message symbol $m \in Z_4$ will add the all 1's vector (modulo 4) to the codeword **c** and that incrementing by 1 (modulo 4) in the QPSK mapping (Table 2.1) corresponds to rotation by 90° (counterclockwise). This rotational invariance is exploited in the standard by using a differential encoding method that involves "precoding" at the transmitter[2]

$$\tilde{m}_k = m_k + \tilde{m}_{k-1} \quad \text{(modulo 4)}$$

and "differential" decoding at the receiver

$$m_k = \tilde{m}_k - \tilde{m}_{k-1} \quad \text{(modulo 4)}$$

(the sliding window nature of the differential decoder limits error propagation).

The 1 Mbit/s mode is defined by using a repetition generator matrix

$$G = [2, 2, 2, 2, 2, 2, 2, 2, 2, 2, 2]$$

which incorporates a binary message symbol, $m \in Z_2 = \{0, 1\}$, and produces a BPSK signal, $x_j \in \{+1 + i, -1 - i\}$. This produces a code that is 180° rotationally invariant.

$\boxed{Example}$ *Barker*

To encode $[m_1 = 1, m_2 = 0, m_3 = 2, m_4 = 3, \ldots]$ the precoder would produce (taking $\tilde{m}_0 = 0$)

$$[\tilde{m}_1 = 1, \tilde{m}_2 = 1, \tilde{m}_3 = 3, \tilde{m}_4 = 2, \ldots].$$

This would be encoded into the Barker codewords according to Eq. (2.1)

$$[31331333111, 31331333111, 13113111333, 02002000222, \ldots]$$

and then translated to the QPSK symbols as in Eq. (2.2)

$$[1 - i, -1 + i, 1 - i, 1 - i, -1 + i, 1 - i, 1 - i, 1 - i, -1 + i, -1 + i, -1 + i,$$
$$1 - i, -1 + i, 1 - i, 1 - i, -1 + i, 1 - i, 1 - i, 1 - i, -1 + i, -1 + i, -1 + i,$$
$$-1 + i, 1 - i, -1 + i, -1 + i, 1 - i, -1 + i, -1 + i, -1 + i, 1 - i, 1 - i, 1 - i,$$
$$1 + i, -1 - i, 1 + i, 1 + i, -1 - i, 1 + i, 1 + i, 1 + i, -1 - i, -1 - i, -1 - i,$$
$$\ldots]$$

according to Table 2.1.                                                    □

---

[2]The precoded symbol at time $k$, $\tilde{m}_k$, is used in the encoding Eq. (2.1).

In any communications system, the reliability of transmission can be improved with a corresponding reduction in transmission rate. For example, by sending a given signal $n$ times, an energy gain factor of $n$ [or $10 \cdot \log_{10}(n)$ dB] can be achieved in signal-to-noise ratio (SNR) at the expense of a factor of $1/n$ in rate (because the same signal is transmitted $n$ times). However, coding theory predicts that for a given reduction in rate $R$, the improvement in SNR can be greater than $1/R$. An improvement in excess of the simple "repeation gain" is commonly known as "coding gain."

The *minimum squared distance* (MSD) of QPSK is $2E_s$ (where $E_s$ is the average symbol energy), whereas BPSK has an MSD of $4E_s$. Both the 1 and 2 Mbit/s transmissions schemes show an energy improvement in minimum distance squared, at the cost of rate. In the case of 2 Mbit/s, the minimum distance squared is $22E_s$ because the Barker encoder has a repeation effect of length 11. This results in an energy gain of $11 = 10.41$ dB over uncoded QPSK with a factor of $1/11$ in the data rate. From a coding gain perspective, there is no coding gain *w.r.t.* QPSK because the minimum distance squared normalized by the data rate is the same as QPSK. The *asymptotic coding gain* (ACG) of a coded system (C) relative to an uncoded system (U) is defined as the ratio

$$ \mathrm{ACG} = \frac{d^2_{\min}(C) \cdot R(C)/E_s(C)}{d^2_{\min}(U) \cdot R(U)/E_s(U)}. $$

In the 2 Mbit/s case, $d^2_{\min}(C)/E_s(C) = 22$ and $R(C) = 2/11$ (bits/symbol), whereas for uncoded QPSK, $d^2_{\min}(U)/E_s(U) = 2$ and $R(U) = 2$; in this case ACG $= 1 = 0$ dB. Similarly in the 1 Mbit/s case, which uses BPSK, there is an energy gain of $2 \cdot 11 = 22 = 13.42$ dB (over QPSK) but 0 dB of coding gain because the data rate factor is $1/22$ of uncoded QPSK.

### 2.3.3   The "High-Rate" Standards: *The Present*

The standard calls for two choices of coding, each involving a "symbol rate" of 11 MHz and data rates of 5.5 and 11 Mbit/s. One code uses a short blocklength code, known as "CCK" that codes over 8 QPSK symbols and the other choice incorporates 64-state, Packet Binary Convolutional Coding (PBCC). The main difference between the two involves the much larger coding gain of the PBCC over CCK at a cost of computation at the receiver.

*2.3.3.1   CCK 5.5 and 11 Mbit/s*   The *Complementary Code Keying* (CCK) code can be considered as a block code generalization of the low-rate Barker code. For CCK-11, the code is an ($n = 8$, $k = 4$) linear block code over $Z_4$. At the 11 Mbit/s rate, 8 bits (4-$Z_4$ symbols) of information is encoded via the $k \times n = 4 \times 8$ CCK-11 *generator matrix*

$$G = \begin{bmatrix} 1 & 1 & 1 & 1 & 1 & 1 & 1 & 1 \\ 1 & 1 & 1 & 1 & 0 & 0 & 0 & 0 \\ 1 & 1 & 0 & 0 & 1 & 1 & 0 & 0 \\ 1 & 0 & 1 & 0 & 1 & 0 & 1 & 0 \end{bmatrix}$$

using the matrix equation

$$\mathbf{c} = \mathbf{m} \cdot G + \mathbf{b} = [c_1, c_2, c_3, c_4, c_5, c_6, c_7, c_8] \quad (\text{modulo } 4) \qquad (2.3)$$

In this case, the length 8 *cover vector* is given by

$$\mathbf{b} = [0, 0, 0, 2, 0, 0, 2, 0]$$

and the message vector, $\mathbf{m} = [m_1, m_2, m_3, m_4]$, $m_j \in Z_4$, represents 8 bits of information. Applying the QPSK mapping shown in Table 2.1 produces the signal vector

$$\mathbf{c} \rightarrow \mathbf{x} = [x_1, x_2, x_3, x_4, x_5, x_6, x_7, x_8].$$

At the 5.5 Mbit/s rate, 4 bits of information is encoded via the $k \times n = 3 \times 8$ CCK-5.5 *generator matrix*

$$G = \begin{bmatrix} 1 & 1 & 1 & 1 & 1 & 1 & 1 & 1 \\ 2 & 2 & 2 & 2 & 0 & 0 & 0 & 0 \\ 2 & 0 & 2 & 0 & 2 & 0 & 2 & 0 \end{bmatrix}$$

using the matrix equation

$$\mathbf{c} = \mathbf{m} \cdot G + \mathbf{b} = [c_1, c_2, c_3, c_4, c_5, c_6, c_7, c_8] \quad (\text{modulo } 4)$$

In this case, the length 8 cover vector is given by

$$\mathbf{b} = [1, 0, 1, 2, 1, 0, 3, 0]$$

and the message vector, $\mathbf{m} = [m_1, m_2, m_3]$, $m_1 \in Z_4$, $m_2 \in Z_2$, $m_3 \in Z_2$, represents 4 bits of information.

The CCK code is rotationally invariant because the first row of the generator matrix $G$ is the all-1's vector. This implies that a rotation by a multiple of 90° at the receiver will affect only the first symbol $m_1$ of the message vector. This is exploited in the standard by differential encoding/decoding on the first symbol $m_1$, using the same method as in the low-rate case.

| Example | CCK-11

The encoding of the 8 bits $[m_1 = 1, m_2 = 0, m_3 = 2, m_4 = 3]$ (ignoring the precoding function on $m_1$) produces the CCK codeword according to Eq. (2.3)

$$1 \cdot [11111111] + 0 \cdot [11110000] + 2 \cdot [11001100] + 2 \cdot [10101010]$$

$$\mathbf{c} =$$

$$+ [10121030]$$

$$= [11111111] + [00000000] + [22002200] + [20202020] + [10121030]$$

$$= [23032321]$$

and then translated to the QPSK symbols as in Eq. (2.2)

$$[-1 - i, 1 - i, 1 + i, 1 - i, -1 - i, 1 - i, -1 - i, -1 + i]$$

according to Table 2.1.                                                  □

The minimum distance squared of the 11 Mbit/s CCK code is $8E_s$; two code-words at minimum distance are generated by the messages $\mathbf{m}_1 = [0\ 0\ 0\ 0]$

$$\mathbf{c}_1 = [0\ \ 0\ \ 0\ \ 2\ \ 0\ \ 0\ \ 2\ \ 0] \rightarrow$$

$$\mathbf{x}_1 = [+1 + i\ \ +1 + i\ \ +1 + i\ \ -1 - i\ \ +1 + i\ \ +1 + i\ \ -1 - i\ \ +1 + i]$$

and $\mathbf{m}_2 = [0001]$,

$$\mathbf{c}_1 = [1\ \ 0\ \ 1\ \ 2\ \ 1\ \ 0\ \ 3\ \ 0] \rightarrow$$

$$\mathbf{x}_1 = [-1 + i\ \ +1 + i\ \ -1 + i\ \ -1 - i\ \ -1 + i\ \ +1 + i\ \ +1 - i\ \ +1 + i]$$

for example. The minimum distance squared of the 5.5 Mbit/s CCK code is $16E_s$. It is interesting to note that a 6.875 Mbit/s CCK code, with the same minimum distance of $16E_s$, is possible by using the generator

$$G = \begin{bmatrix} 1 & 1 & 1 & 1 & 1 & 1 & 1 & 1 \\ 2 & 2 & 2 & 2 & 0 & 0 & 0 & 0 \\ 2 & 2 & 0 & 0 & 2 & 2 & 0 & 0 \\ 2 & 0 & 2 & 0 & 2 & 0 & 2 & 0 \end{bmatrix}$$

This code is *not* part of the standard.

The asymptotic coding gain for CCK is 3 dB (ACG = 2) over uncoded QPSK. This follows the fact that the code rate is 1/2 and the minimum distance is 4; the product is 2. However, the practical coding gain is about 2 dB (as shown in Section 2.4.2). The reduction in coding gain from the asymptote is due to the number of "nearest neighbors" at the minimum distance as shown in Table 2.2. This table shows that at the minimum distance of the code ($8E_s$ for CCK-11 and $16E_s$ for CCK-5.5/6.875) there are 24/14/30 codewords. This large number of nearest neighbors (compared with 2 nearest neighbors for the 2 Mbit/s Barker) accounts for the 1-dB reduction in practical coding gain.

Because the CCK codes are affine translations of linear block codes, the codewords can be compactly described in terms of a trellis with $n = 8$ sections as shown in Figure 2.6. A description of the generation of the trellis for a linear block code is given in Chapter 2 of *Turbo Coding* [6]. In the case of CCK-11, the number of states of the trellis follow a [1, 4, 16, 64, 16, 64, 16, 4, 1] pattern; there are 296 branches in the trellis.

The trellis can be derived from a parity check matrix

$$H = \begin{bmatrix} 0 & 1 & 0 & 3 & 3 & 0 & 1 & 0 \\ 1 & 3 & 3 & 1 & 0 & 0 & 0 & 0 \\ 0 & 0 & 1 & 3 & 3 & 1 & 0 & 0 \\ 0 & 0 & 0 & 0 & 1 & 3 & 3 & 1 \end{bmatrix}$$

a $4 \times 8$ matrix over $Z_4$. Note that on the 1st, 2nd, 3rd and 5th trellis sections, there is 4-way branching and on the 4th, 6th, 7th and 8th trellis sections there is 1-way branching. The trellis for CCK-5.5 has a [1, 4, 8, 8, 4, 8, 8, 4, 1] state pattern with 56 branches and 4-way branching on the 1st trellis section and 2-way branching on the 2nd and 5th trellis sections. A parity check that generates this trellis is given by the $7 \times 8$ matrix

$$H = \begin{bmatrix} 1 & 0 & 3 & 0 & 0 & 0 & 0 & 0 \\ 0 & 1 & 0 & 3 & 0 & 0 & 0 & 0 \\ 0 & 0 & 1 & 3 & 3 & 1 & 0 & 0 \\ 0 & 0 & 0 & 0 & 1 & 0 & 3 & 0 \\ 0 & 0 & 0 & 0 & 0 & 1 & 0 & 3 \\ 0 & 0 & 1 & 1 & 1 & 1 & 0 & 0 \\ 1 & 2 & 1 & 0 & 0 & 0 & 0 & 0 \end{bmatrix}$$

**Table 2.2. CCK Weight Distribution.**

| Wt/$2E_s$: | 0 | 4 | 6 | 8 | 10 | 12 | 16 |
|---|---|---|---|---|---|---|---|
| Number (CCK-11): | 1 | 24 | 16 | 174 | 16 | 24 | 1 |
| Number (CCK-5.5): | 1 | | | 14 | | | 1 |
| Number (CCK-6.875): | 1 | | | 30 | | | 1 |

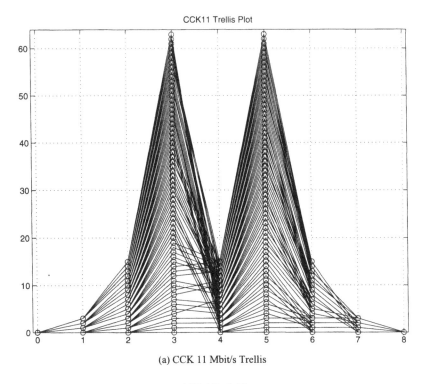

(a) CCK 11 Mbit/s Trellis

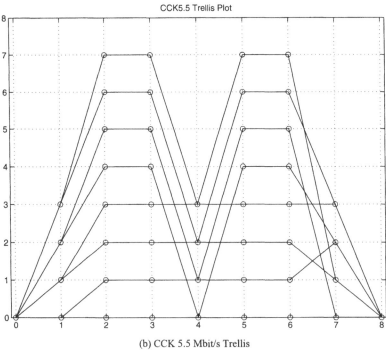

(b) CCK 5.5 Mbit/s Trellis

**Figure 2.6.** The trellis for CCK.

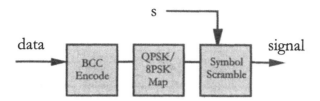

**Figure 2.7.** Packet Binary Convolutional Coding.

### 2.3.3.2  PBCC 5.5 and 11 Mbit/s

*2.3.3.2  PBCC 5.5 and 11 Mbit/s*  The IEEE 802.11b standard specifies an optional choice of coding and modulation and is considered the "high-performance" mode for 11 and 5.5 Mbit/s transmission. The optional mode, termed *Packet Binary Convolutional Coding* (PBCC), involves a BCC combined with a symbol scrambling method as shown in Figure 2.7. This structure is also used for the higher-rate (22 Mbit/s) encoding described in Section 2.3.4.1.

The 802.11b PBCC mode (11 and 5.5 Mbit/s) uses a $1 \times 2$ generator matrix over $Z_2[D]$, the set of polynomials (in the variable $D$) with binary coefficients:

$$G = [D + D^2 + D^5 \quad 1 + D^2 + D^3 + D^4 + D^5 + D^6] \qquad (2.4)$$

as shown in Figure 2.8(a) (in octal notation $G = [46, 175]$). For 11 Mbit/s operation, this 64-state encoder is followed by a mapping onto QPSK modulation directly as specified in Table 2.3.[3] For 5.5 Mbit/s, the two binary outputs are bit serialized and mapped onto BPSK.

The last operation of the encoder is the "symbol scrambling." A specified 256-bit periodic binary sequence is used to control the symbol scrambler. When the binary "s" value into the symbol scrambler is "0," the QPSK/BPSK symbol out of the symbol mapper is sent directly, while an $s = 1$ tells the symbol scrambler to rotate the mapped symbol by 90° (counterclockwise) as shown in Figure 2.9.

Generation of the period 256 "s" sequence can be described in a two-step process. First, a balanced (half 0s, half 1s) vector of length 16 is given by

$$\mathbf{u} = [u_0, u_1, \ldots u_{15}] = [0011001110001011]$$

This vector is repeatedly concatenated with an order 3 circular rotation of the previous vector

$$\mathbf{s} = \mathbf{u} \circ \sigma^3(\mathbf{u}) \circ \sigma^6(\mathbf{u}) \circ \sigma^9(\mathbf{u}) \circ \ldots$$

---

[3]The mapping given in Table 2.3 does not map "Hamming distance" to "Euclidean distance." An equivalent encoder would map $00 \rightarrow +1 + i$, $01 \rightarrow -1 + i$, $11 \rightarrow -1 - i$, $10 \rightarrow +1 - i$ and use a BCC generator $G = [133, 175]$.

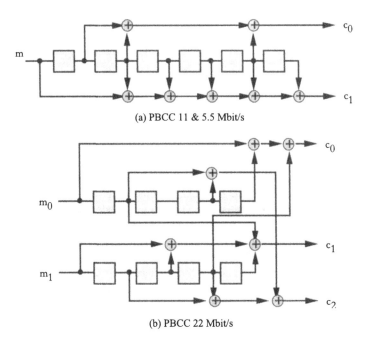

(a) PBCC 11 & 5.5 Mbit/s

(b) PBCC 22 Mbit/s

**Figure 2.8.** The binary convolutional encoders.

where the 3rd circular shifted is defined by

$$\sigma(\mathbf{u}) \equiv [u_1, u_2, u_3, \ldots, u_{13}, u_{14}, u_{15}, u_0] = [0110011100010110],$$

$$\sigma^2(\mathbf{u}) \equiv [u_2, u_3, u_4, \ldots, u_{14}, u_{15}, u_0, u_1] = [1100111000101100],$$

$$\sigma^3(\mathbf{u}) \equiv [u_3, u_4, u_5, \ldots, u_{15}, u_0, u_1, u_2] = [1001110001011001].$$

The "o" symbol represents the concatenation operator, thus the 16 bits in **u** is followed by the 16 bits of $\sigma^3(\mathbf{u})$, followed by the 16 bits of $\sigma^6(\mathbf{u})$, etc. The chosen vec-

**Table 2.3.  QPSK/BPSK Mapping (PBCC)**

| Code Label $c_1 c_0$ | QPSK Signal $x_i$ | Code Label $c_1$ | BPSK Signal $\&x_i$ |
|---|---|---|---|
| 00 | $+1 + i$ | 0 | $+1 + i$ |
| 01 | $-1 + i$ | 1 | $-1 - i$ |
| 10 | $-1 - i$ | | |
| 11 | $+1 - i$ | | |

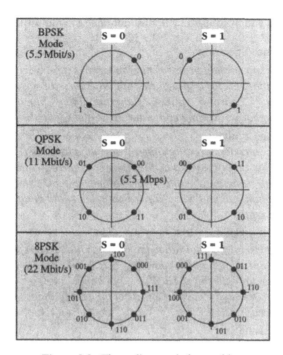

**Figure 2.9.** The coding symbol scrambler.

tor **u**, combined with the fact that 3 and 16 are relatively prime, means that $\sigma^{3m}(\mathbf{u})$ are distinct for $m = 0, 1, \ldots, 15$ and $\sigma^{48}(\mathbf{u}) = \mathbf{u}$. Thus this method of symbol scrambler sequence generation has a period of $16 \times 16 = 256$ bits.

[Example] *PBCC-11*

The encoding of the message bits $[m_1, m_2, m_3, m_4, m_5, m_6, m_7, m_8, \ldots] = [1, 0, 1, 1, 1, 0, 1, 0, \ldots]$ produces the BCC codeword according to Eq. (2.4)

$$\mathbf{c} = [10, 01, 01, 01, 10, 11, 11, 10, \ldots].$$

The BCC outputs are translated to the QPSK symbols using Table 2.3

$$[-1 - i, -1 + i, -1 + i, -1 + i, -1 - i, 1 - i, 1 - i, -1 - i, \ldots]$$

and selectively rotated according to $s = [00110011 \ldots]$

$$[-1 - i, -1 + i, -1 - i, -1 - i, -1 - i, 1 - i, 1 + i, 1 - i, \ldots]$$

as described in Figure 2.9.                                                    □

The characteristics and benefits of symbol scrambling are multifold—

- Signal Distance Spectrum

  *The distance spectrum of the transmission signal set is invariant to the scrambling operation. This is a consequence of the distance-preserving nature of the 90° rotation [7]. However, unlike a "data scrambling" function (a one-to-one function), symbol scrambling does alter the signal set in beneficial ways.*

- Time Varying Coding

  *Typical BCC encoders produce time-invariant codewords. This means that a time-shifted version of a valid code sequence is also a valid code sequence. The periodic scrambling, with a long 256 period, makes the code sequences appear aperiodic (actually they are periodic, but with a long period). This effect can be useful.*

- Interference Rejection

  *When an interfering signal is added to a transmitted packet, it is helpful if the interferer is not a legitimate codeword. This is the case for an aperiodic encoding. Thus, for interferers such as co-channel interference or unmodeled multipath distortion, the adverse effects of the interfering signal can be significantly reduced.*

- Tone Suppression

  *Time-invariant convolutional coding can generate codewords with unwanted spectral characteristics. For example, the all-0's message will produce an all-0's codeword that, without the symbol scrambler, produces a constant transmission signal. A similar effect will occur if a (small) periodic message is encoded into a periodic codeword. The symbol scrambler removes this signaling possibility, ensuring that signals with poor spectral characteristics are never transmitted.*

The BCC encoder selected for the PBCC-11 code involves a trade-off between optimal additive white Gaussian noise (AWGN) performance and tolerance to multipath and other forms of interference. The NASA standard 64-state code (with generator $G = [133, 171]$) [8] is optimized to maximize the Euclidean free distance $d_{\text{free}}$ = 10; the Euclidean free distance of the PBCC-11 code is $d_{\text{free}}$ = 9. The Euclidean distance spectrum of these two codes is shown in Table 2.4. These data show that the PBCC-11 code has only one error event of weight 9 and 6 error events at distance 10, whereas the NASA code has 11 error events of weight 10. These facts ex-

**Table 2.4. PBCC-11 Euclidean Weight Distribution.**

| Wt/2$E_s$: | 0 | 9 | 10 | 11 | 12 | 13 | 14 | 15 | 16 | . . . |
|---|---|---|---|---|---|---|---|---|---|---|
| Number (PBCC-11): | 1 | 1 | 6 | 11 | 12 | 45 | 117 | 259 | 629 | . . . |
| Number (NASA): | 1 | 0 | 11 | 0 | 38 | 0 | 193 | 0 | 1,331 | . . . |

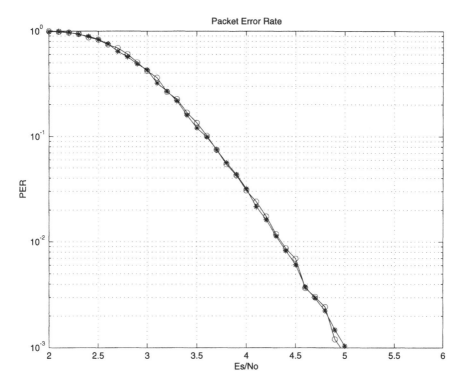

**Figure 2.10.** Comparison of PBCC-11 and the NASA Code (1,000-byte packets).

plain why the PBCC-11 has only an insignificant loss in SNR, if any, over the AWGN channel (a very small fraction of a dB) as shown in Figure 2.10. The asymptotic coding gain for PBCC-11 is 6.5 dB (ACG = 4.5) over uncoded QPSK. The practical coding gain is about 5.5 dB (as shown in Section 2.4.2). It is interesting to note that the NASA code has an ACG = 5 = 6.9 dB (0.4 dB higher), yet the practical gain is the same 5.5 dB.

Table 2.5 shows a definite advantage for the PBCC-11 code. In this table the symbol or "Hamming" weight distributions of the two codes are compared. It can be seen here that an error event for the PBCC-11 code will span at least

**Table 2.5. PBCC-11 Symbol (Hamming) Weight Distribution**

| Symbol Weight: | 0 | 6 | 7 | 8 | 9 | 10 | 11 | 12 | . . . |
|---|---|---|---|---|---|---|---|---|---|
| Number (PBCC-11): | 1 | 0 | 6 | 8 | 20 | 78 | 204 | 639 | . . . |
| Number (NASA): | 1 | 1 | 4 | 10 | 21 | 66 | 222 | 617 | . . . |

seven QPSK symbols whereas the NASA code has an error event that spans only six QPSK symbols. It was this trade-off between Euclidean distance and symbol distance that led to the selection of the PBCC-11 for the IEEE 802.11b standard.

### 2.3.4    The "Higher-Rate" Standards: *The Future*

The Alantro/TI proposal increases the data rate of the IEEE 802.11b standard in a backward compatible way.

When the engineering team at Alantro started the higher rate project, the following constraints were of main concern—

- Interoperability with IEEE 802.11b networks
  *Introduction of higher rate transmission in an existing network is a prime requirement.*
- Translate coding gain advantage to "double the data rate"
  *22 Mbit/s*
- Compatibility with IEEE 802.11b radios
  *8-PSK, 11 MHz symbol rate, short preamble*
- Operate in the same environment as CCK-11
  *64 state code → 256 state code; a good engineering solution: cost versus performance*
- Satisfy FCC Requirements
  *Same spectral and temporal signal characteristics as IEEE 802.11b; noise and interference tolerance comparable to CCK-11.*

*2.3.4.1  PBCC 22 Mbit/s*    The high-rate case (22 Mbit/s) has a $2 \times 3$ generator matrix over $Z_2[D]$:

$$G = \begin{bmatrix} 1 + D^4 & D & D + D^3 \\ D^3 & 1 + D^2 + D^4 & D + D^3 \end{bmatrix}$$

(in octal notation $G = [21, 2, 12; 10, 25, 12]$)

**Table 2.6. 8PSK Mapping.**

| Code Label $c_2 c_1 c_0$ | 8PSK Signal $y_i$ | Digital-8PSK $x_i$ | $c_2 c_1 c_0$ | $y_i$ | $x_i$ |
|---|---|---|---|---|---|
| 000 | $+1 + i$ | $+5 + 5i$ | 100 | $\sqrt{2}$ | $+7i$ |
| 001 | $-1 + i$ | $-5 + 5i$ | 101 | $\sqrt{2}i$ | $-7$ |
| 010 | $-1 - i$ | $-5 - 5i$ | 110 | $-\sqrt{2}$ | $-7i$ |
| 011 | $+1 - i$ | $+5 - 5i$ | 111 | $-\sqrt{2}i$ | $+7$ |

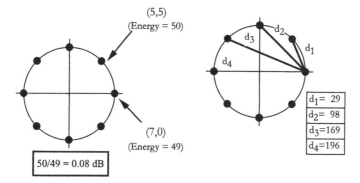

**Figure 2.11**  Digital-8PSK.

A 1 × 3 parity check matrix:

$$H = \begin{matrix} [D + D^2 + D^4 + D^7 & D + D^3 + D^4 + D^5 + D^6 + D^7 \cdots \\ & \cdots 1 + D^2 + D^4 + D^6 + D^8] \end{matrix}$$

(In octal notation $H = [226, 372, 525]$). This BCC encoding function is combined with the "Digital-8PSK" signal mapping shown in Table 2.6 to produce a coded eight-level modulation signal.

This coded modulation was discovered via computer search with a bounding technique illustrated in Figure 2.11 and Table 2.7. The weight values in the table provide a lower bound on the distance between points in the signal constellation. If $(c_2 c_1 c_0)$ and $(c_2' c_1' c_0')$ are the labels of two points, then

$$\|x_i(c_2 c_1 c_0) - x_i(c_2' c_1' c_0')\|^2 \ge w(c_2 \oplus_2 c_2', c_1 \oplus_2 c_1', c_0 \oplus_2 c_0') \qquad (2.5)$$

where the operation $\oplus_2$ is modulo 2 addition. Furthermore, this bound is tight [i.e., for each value of $w$ there is a pair of labels that achieve equality in Eq. (2.5)]. The

**Table 2.7.  Digital-8PSK Weight Bound.**

| Code Label $c_2 c_1 c_0$ | Weight $w(c_2 c_1 c_0)$ | Code Label $c_2 c_1 c_0$ | Weight $w(c_2 c_1 c_0)$ |
|---|---|---|---|
| 000 | 0 | 100 | 29 |
| 001 | 98 | 101 | 29 |
| 010 | 196 | 110 | 169 |
| 011 | 98 | 111 | 29 |

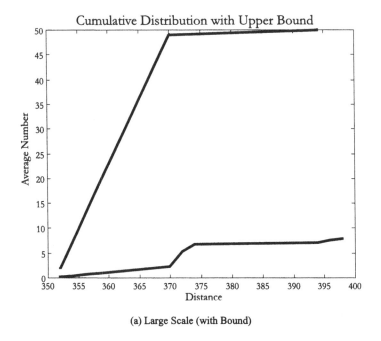

(a) Large Scale (with Bound)

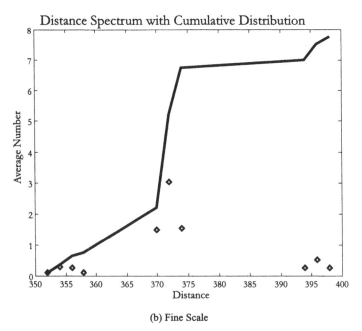

(b) Fine Scale

**Figure 2.12.** The distance spectrum for PBCC-22.

**Table 2.8. PBCC-22 Weight Distribution Bound.**

| Wt/2E$_s$:<br>99 · Wt/2E$_s$: | 3.56<br>352 | 3.74<br>370 | 3.98<br>394 | 4.14<br>410 | 4.32<br>428 | 4.55<br>450 | . . .<br>. . . |
|---|---|---|---|---|---|---|---|
| Number: | 2 | 47 | 1 | 53 | 437 | 12 | . . . |

use of this weight function to compare the accumulated distance on a pair of sequences is the basis for the computer search.

Figure 2.12(a) shows a plot of the distance spectrum of the PBCC-22 code as well as the bound that was used in the search. One can see that the bound predicts the free distance of the code $d_{free} = 352$ but overestimates the growth in nearest neighbors. Figure 2.12(b) shows the average nearest neighbor growth near the free distance of the code, the data for these graphs are presented in Tables 2.8 and 2.9.

## 2.4    Performance

### 2.4.1    Spectrum

The power spectrum for all the transmission modes are essentially the same with a small deviation for the original 1 and 2 Mbit/s modes. To obtain the theoretical power spectral density for a complex waveform of the form

$$x(t) = \sum_{k=-\infty}^{\infty} A_k p(t - kT_s) \tag{2.6}$$

**Table 2.9. PBCC-22 Average Weight Distribution.**

| Wt/2E$_s$:<br>99 · Wt/2E$_s$: | 3.56<br>352 | 3.58<br>354 | 3.60<br>356 | 3.62<br>358 | 3.74<br>370 | 3.76<br>372 | 3.78<br>374 | 3.98<br>394 |
|---|---|---|---|---|---|---|---|---|
| Ave. Number: | 0.0913 | 0.2783 | 0.2677 | 0.0927 | 1.479 | 3.017 | 1.528 | .2497 |

| Wt/2E$_s$:<br>99 · Wt/2E$_s$: | 4.00<br>396 | 4.02<br>398 | 4.14<br>410 | 4.16<br>412 | 4.18<br>414 | 4.20<br>416 | 4.32<br>428 | 4.34<br>430 |
|---|---|---|---|---|---|---|---|---|
| Ave. Number: | 0.5 | .2503 | 1.293 | 2.786 | 2.796 | 1.327 | 3.843 | 7.786 |

| Wt/2E$_s$:<br>99 · Wt/2E$_s$: | 4.36<br>432 | 4.55<br>450 | 4.57<br>452 | 4.59<br>454 | 4.61<br>456 | 4.63<br>458 | . . .<br>. . . | |
|---|---|---|---|---|---|---|---|---|
| Ave. Number: | 3.933 | 0.282 | 1.894 | 3.267 | 1.848 | 0.2693 | . . . | |

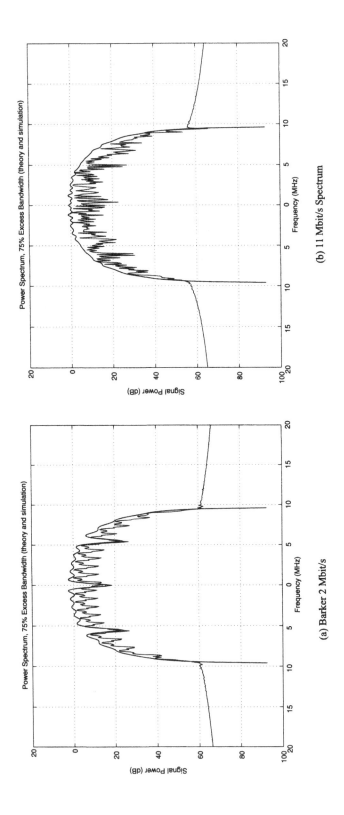

(a) Barker 2 Mbit/s

(b) 11 Mbit/s Spectrum

46

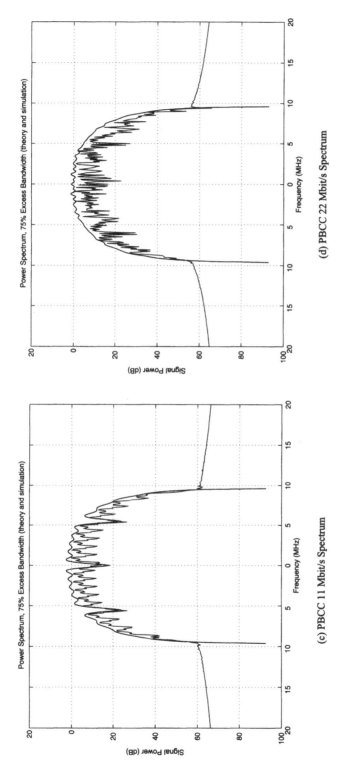

**Figure 2.13.** The power spectrum of various codings.

47

where $A_k$ is a random symbol sequence, $p(t)$ is the pulse shape, and $T_s$ is the symbol period, the power spectral density is given by the formula

$$S_x(f) = \frac{1}{T_s}|P(f)|^2 S_A(fT_s)$$    (2.7)

where $P(f)$ is the Fourier transform of the pulse shape and

$$S_A(f) = \sum_l R_A(l)e^{i2\pi lf}$$

is the discrete Fourier transform of the autocorrelation function for the symbol sequence. The fact that Eq. (2.7) is the product of two terms shows that the effect of a nontrivial symbol autocorrelation is to modulate the shape of the pulse spectrum. This formula is the basis of the theoretical curves offered in Figure 2.13 and shows very good agreement with experimental results.

Figure 2.14 shows the autocorrelation for the Barker encoder[4] described in Eq. 2.1 on page 30. This nontrivial autocorrelation results in small "ripples" in the power spectral density as observed in both the theoretical and experimental power spectral results shown in Figure 2.13(a). Both the CCK code described in Section 2.3.3.1 on page 32 and the PBCC codes described in Section 2.3.3.2 on page 37 offer "white" symbol sequences. This is verified in Figures 2.13(b), 2.13(c), and 2.13(d).

### 2.4.2    AWGN Performance

The performance of the various combinations of modeling and modulation is presented in Figures 2.15–2.18. In Figure 2.15, the *Bit Error Rate* (BER) of the various choices is shown as a function of the received signal-to-noise ratio $E_s/N_o$. Figure 2.16 shows the *Packet Error Rate* (PER), for 1,000-byte (8,000 bits) packets as a function of the received signal-to-noise ratio $E_s/N_o$. Figure 2.17 shows the PER as a function of the energy per bit-to-noise ratio $E_b/N_o$; these curves are useful for computing and comparing the practical coding gains of the systems. Finally, Figure 2.18 shows the PER as a function of the received signal to noise ratio $E_s/N_o$ for the 22-Mbit/s system with the multipath receiver that is the basis of the Alantro/TI baseband receiver product. The multipath is modeled with a method developed by the IEEE 802.11 committee that indexes the multipath by a factor known as the "delay spread" [9]. In this model, an increase in delay spread corresponds to a more severe multipath environment.

---

[4]The symbol sequence $A_k$ is defined with random data encoded according to Eq. (2.1) and (2.2) on page 51 and a uniformly distributed phase $A_k = x_{k-N}$, $0 \le N < 11$. The DFT of the autocorrelation of the Barker sequence

$$S_A(f) = 1 - \frac{2}{11} \sum_{l=1}^{5} \cos(4\pi lf)$$

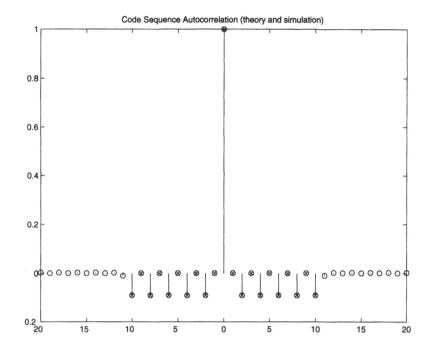

**Figure 2.14.** Barker autocorrelation.

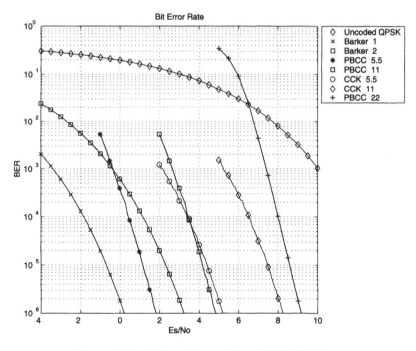

**Figure 2.15.** Bit Error Rate vs. Channel SNR ($E_s/N_o$).

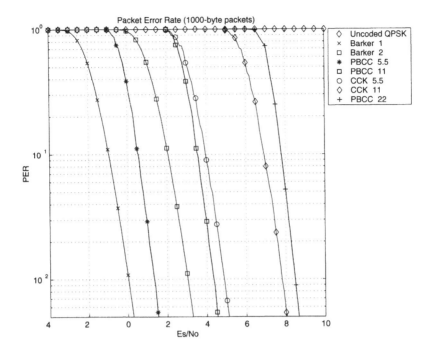

**Figure 2.16.** Packet Error Rate vs. Channel SNR ($E_s/N_o$).

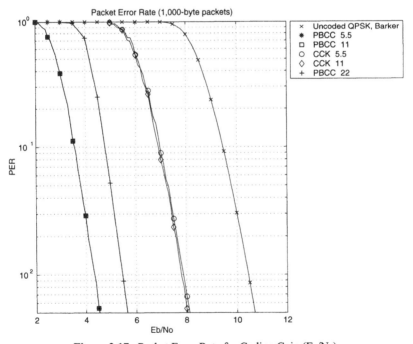

**Figure 2.17.** Packet Error Rate for Coding Gain ($E_b/N_o$).

50

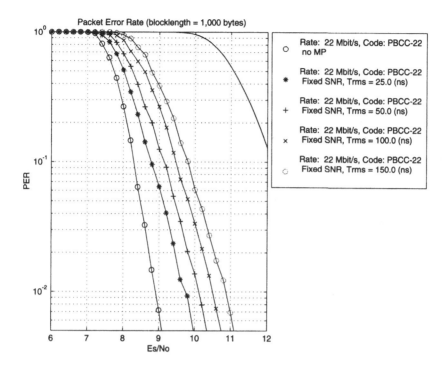

**Figure 2.18.** Packet Error Rate in Multipath (22 Mbit/s).

### 2.4.3    Computational Complexity

A comparison of the computational requirements to decode all the high-rate modes is given in Table 2.10. This table shows the number of basic computations required to perform an optimal decoding in AWGN with the Viterbi algorithm [8]. Note that these results do not consider the cost of dealing with the prevailing issue in wireless Ethernet, multipath. Thus these results, which are useful for a raw comparison of

**Table 2.10.  Trellis Complexity with Viterbi Decoding Compared.**

| Code | Branches per Information Bit | Megabranches per Second |
|------|------------------------------|-------------------------|
| CCK5.5 | 14 | 154 |
| CCK11 | 37 | 407 |
| PBCC5.5 | 128 | 704 |
| PBCC11 | 128 | 1,408 |
| PBCC22 | 1,024 | 11,264 |

the various coding schemes, do not give a complete picture of complexity required to implement a wireless Ethernet baseband processor.

## 2.5    Spread Spectrum Transmission

In wireless communications, such as IEEE 802.11b and other shared media systems, information is often encoded with spread spectrum signaling methods. The *spectral efficiency* of a digital transmission system is defined as the ratio of the user *data rate* (in bits/second) to the *bandwidth* (in Hertz) of the power spectral density (suitably defined) of the ensemble of transmission signals. In a very thought-provoking chapter [10], Jim Massey considered an information theory definition of spread spectrum and studied some of the consequences of his view. This definition is described in some detail in Section 2.5.1. For example, Massey's definition implies that in spread spectrum signaling systems, the spectral efficiency will be low. He also showed, via examples, that the converse is not true: a low spectral efficiency does not imply a spread spectrum signal set.

Massey demonstrated that in systems with a low spectral efficiency, the use of spread spectrum is a reasonable means of communications that has only a modest, acceptable loss in Shannon capacity. He also showed that in high-spectral-efficiency systems, mathematically precise notions of spread spectrum imply a very significant, uneconomic loss in capacity. In the Massey framework, if the spectral efficiency is not a small fraction of 1, spread spectrum is not practical.

This view is in contrast to the view of the U. S. Federal Communications Commission (FCC), which uses a more pragmatic definition of spread spectrum. The FCC defines direct sequence spread spectrum in a much less restrictive way. According to the FCC:

> Direct Sequence Systems "*A spread spectrum system in which the carrier has been modulated by a high speed spreading code and an information data stream. The high speed code sequence dominates the 'modulating function' and is the direct cause of the wide spreading of the transmitted signal.*"

and

> Spread Spectrum Systems "*A spread spectrum system is an information bearing communications system in which: (1) Information is conveyed by modulation of a carrier by some conventional means, (2) the bandwidth is deliberately widened by means of a spreading function over that which would be needed to transmit the information alone. (In some spread spectrum systems, a portion of the information being conveyed by the system may be contained in the spreading function.)*"

It is interesting to observe that it is the last parenthetical element that differentiates the strict requirements of Massey's definition and that the FCC rules that have opened the door to higher spectral efficiencies and user data rates in the "ISM" 2.4-GHz band.[5] Without flexibility and pragmatism on the part of the FCC, a more technically strict definition such as Massey's would have prevented the widespread success of the IEEE 802.11b standard. It has been indicated by the FCC that as the standardization process continues to make progress in the development of higher-performing wireless Ethernets, regulators will continue to support the needs of the industry and consumers.

In the process of significantly increasing the data rate, the spread spectrum nature of the signal, in the narrow sense of Massey, is sacrificed. However, the flexible FCC definition allowed the FCC to certify the existing IEEE 802.11b 11 Mbit/s systems under direct sequence spread spectrum rules. This practical approach to regulation is based on the fact that, as an interferer, the high-rate IEEE 802.11b signals are the same as the classic low-rate Barker signals. This is true both in the frequency characteristics, as shown in the power spectral density (Fig. 2.13), and in the time domain or the temporal characteristics of the transmitted signals.

Furthermore, the IEEE 802.11 specifies three disjoint frequency bands for wireless Ethernet systems This means that the legacy 2 Mbit/s systems send a total of 6 Mbit/s in the entire ISM band whereas the 11 Mbit/s systems supply 33 Mbit/s in the band; the 22 Mbit/s systems double the total capacity to 66 Mbit/s.

Radio spectrum is a rare and valuable resource, and it is the responsibility of the FCC to ensure that the resource is used for the public good and in an efficient way. One compelling issue is the demand from the public for higher performance data transmission. Another important issue is the need to avoid the introduction of new signals with spectral and temporal characteristics that were formerly disallowed under the existing rules. Such a change threatens the large base of current products that were built under existing rules with interference that was not previously allowed or anticipated; from a fairness position, this is unjust.

With the huge success of the IEEE 802.11b standard, one can see the wisdom of the FCC. It is anticipated that future regulations will continue to satisfy the demands for higher performance while maintaining a level playing field. The beauty of the PBCC-22 modulation approach is that the data rate is doubled while backward compatibility with existing networks is maintained by using a signal with the same interference characteristic as the existing signal sets. The noise immunity or "processing gain" of the system is the same as that of the CCK-11 system. The spread spectrum nature of the new signals, in the sense of Massey, is the same as the existing systems; this is discussed in Section 2.5.1. It is demonstrated that from an information theory viewpoint, the spread spectrum nature of the new signals is identical to that the existing signal sets used in currently deployed networks. Thus, under any reasonable definition, the PBCC-22 and the CCK-11 systems are equally spread spectrum.

---

[5]The "ISM" band is 83.5 MHz wide, using the range 2.4000–2.4835 GHz.

### 2.5.1  Massey's Definition of Spread Spectrum

Massey defined two notions of bandwidth and argued that the indication of spectrum spreading was related to the size of the ratio of the two. The first definition of bandwidth relates to the spectral occupancy of a given signal or a collection of signals. This form of bandwidth, $B_F$, is known as the "Fourier bandwidth" and relates to the span of frequencies occupied by the signal(s). As is often the case in communication theory, the exact numerical value of the Fourier bandwidth for a given signal or set of signals depends on measurement criteria such as "3-dB" bandwidth or 95% power bandwidth, etc. Such required criteria are often needed to define other quantities of interest in communications theory; examples include the definition of signal-to-noise ratio (SNR) and power spectral density. The Fourier bandwidth is directly related to the "Nyquist bandwidth" [11], which relates to periodic sampling of a signal (or sets of signal) and is of fundamental importance in the study of digital signal processing (DSP).

Massey's second notion of bandwidth is related to the fundamental problem of information transmission and is meaningful only to define a collection or a set of signals. Fundamentally, the problem of information transmission is one of signal design and signal detection. Massey logically argues that the definition of spread spectrum should only involve the signal design issue and not signal detection (i.e., the determination of spread spectrum character of a transmission scheme should not change with a change in the receiver).

Signal design involves the creation of a collection of signals used by a transmitter to represent the multitude of messages that the transmitter is trying to convey. In the signal design problem, various parameters are considered to optimize the transmission systems. Such parameters include transmission power, Fourier bandwidth, power spectrum and data rate, and a host of others including the dimensionality of the signal set.

The *data rate* parameter of a signal set relates to the size of the collection or *number* of signals in the signal set; a system transmits at a rate of $R$ bits per second if, over a time interval of length $T$ seconds, the designed signal set defines $2^{RT}$ distinct signals. With such a collection of signals, $k = RT$ bits of information can be transmitted by assigning a correspondence between the list of signals in the signal set and the $2^k$ possible values for a $k$-bit message.

The *dimensionality* of a signal set involves the standard notion of basis as defined in the area of linear algebra. Roughly speaking, the *dimension* of a signal set relates to the *minimum* number of *independent* parameters (i.e., numbers) required to describe the collection of signals.

The second definition of bandwidth, $B_S$, relates to the dimensionality of a signal set and describes the linear complexity of the scheme; a system transmits using a bandwidth of $B_S$ Hz if, over a time interval of length $T$ seconds, the designed signal set has a basis with $B_S T$ elements. Because of the strong relationship between this notion of bandwidth and information theory, Massey called this second definition the "Shannon bandwidth."

Note that the Fourier bandwidth, the Shannon bandwidth, and the data rate are distinct ideas that all describe attributes of a signal set. For example, the spectral efficiency of a system is the ratio of the data rate to the Fourier bandwidth $R/B_F$. Another important parameter is the *spreading ratio* $\rho = B_F/B_S$, which relates the two notions of bandwidth.

The first observation that Massey noted was the theorem that says that the Fourier bandwidth is never less than the Shannon bandwidth, $B_F \geq B_S$. This means that the spreading ratio satisfies the inequality

$$\rho = \frac{B_F}{B_S} \geq 1$$

Furthermore, Massey argued that the spreading ratio is the logical measure of the degree in which a communications system spreads the spectrum. If a given system has a large value for $\rho$, say 10 or 100, then it should be considered a spread spectrum system, and conversely, a system with a spreading ratio $\rho$ near the minimum of 1 would not be labeled a spread spectrum system. It would be debatable whether a system with a spreading ratio of say $\rho = 4$ is spread spectrum or not; this is the "gray" area.

In Shannon's original 1948 paper [12], a famous formula for the capacity of a band-limited additive white Gaussian channel was presented

$$C(P/N_o, B_F) = B_F \log_2\left(1 + \frac{P}{N_o B_F}\right) \text{ bits/second} \qquad (2.8)$$

where $P$ is the signal power, $N_o$ is the white Gaussian noise level, and $B_F$ is the permissible Fourier bandwidth. The interpretation of the Shannon capacity is that reliable transmission is possible, for a given signal-to-noise ratio (SNR) $P/N_o$ and Fourier bandwidth $B_F$, *if and only if* the rate of transmission is no more than the Shannon capacity $C$. In practical terms, the Shannon limit defines an objective data rate goal for a given signaling environment. For the past 54 years, communications engineering has been striving to approach this goal.

If one is to impose the requirement that the transmission system operate with a required spreading ratio of $\rho$, then the formula is modified to be

$$C(P/N_o, B_F, \rho) = \frac{B_F}{\rho} \log_2\left(1 + \frac{P\rho}{N_o B_F}\right) \text{ bits/second} \qquad (2.9)$$

To understand the limitations imposed on the Shannon capacity when spreading is introduced, it is helpful to interpret Eq. (2.9).

First it should be noted that spreading in this sense incurs a loss in capacity; for a fixed SNR and bandwidth, the Shannon capacity monotonically decreases with increasing spreading $\rho$; if $\rho > 1$, $C(P/N_o, B_F) \equiv C(P/N_o, B_F, 1) > C(P/N_o, B_F, \rho)$. However, as noted in Massey's paper, there are often situations in which the loss is small and spreading is reasonable. The modified Shannon formula, Eq. (2.9), involves the product of two terms, the symbol frequency $(B_F/\rho)$ measured in "symbols per second" and the normalized data rate $[\log_2(1 + P\rho/N_oB_F)]$ measured in "bits per symbol." The spectral efficiency of a system, which is the data rate divided by the Fourier bandwidth, is the normalized rate divided by the spreading ratio and a capacity given by the expression $[(1/\rho) \cdot \log_2(1 + P\rho/N_oB_F)]$.

Spreading is useful only when the normalized rate or the spectral efficiency is very small.[6] Because the normalized rate grows with the spreading ratio $\rho$, for a given situation (i.e., SNR and bandwidth), there will be a practical limit on the spreading ratio. For example, obtaining 90% of the Shannon capacity, with a very modest spreading ratio of $\rho = 2$, requires that the normalized rate be less than about 0.3 bits per symbol and a spectral efficiency of less than 0.15 bits-per-second per Hz. Similarly, a system with a spreading ratio of $\rho = 10$ operating with a tiny spectral efficiency of 0.01 bits-per-second per Hz will incur a greater than 20% loss in Shannon capacity from the spreading.

### 2.5.2    Spread Spectrum in Wireless Ethernet

It is interesting to see how Massey's notion of spreading relates to the DSSS wireless Ethernet standard and the higher rate extensions. In terms of the coding level, the Barker systems introduce a nontrivial spreading ratio of $\rho = 11$ (2 Mbit/s) and $\rho = 22$ (1 Mbit/s). All of the high-rate (> 2 Mbit/s) cases have $\rho = 1$, with the exception of PBCC-5.5, which has $\rho = 2$. In practice, the wireless Ethernet signals use a nontrival excess bandwidth pulse shape so that the occupied bandwidth is larger than the 11-MHz symbol rate. A comparison of the spreading ratio for the various choices is given in Table 2.11. It is important to note that, in terms of Massey's spread ratio, all the high-rate systems have the same value (with the exception of PBCC-5.5). Thus, for example, from the viewpoint of information theory, the CCK-11 and the PBCC-22 signals show the same degree of of signal spreading.

In Figure 2.19 the offered data rate and signal-to-noise ratio requirements for the IEEE 802.11b standard and the Alantro 22 Mbit/s extension are displayed. On the $x$-axis is the signal-to-noise ratio defined as the symbol energy-to-noise ratio $E_s/N_o$,[7] whereas the $y$-axis is the data rate of the system assuming the common 11-MHz symbol frequency that is common to the standard. The upper solid curve is the Shannon limit as described by Eq. (2.8), on page 55. The dotted curve shows the Shannon limit assuming a spreading ratio of $\rho = 11$ in Eq. (2.9) (this is the spreading ratio of the 2 Mbit/s Barker system). The individual points on the graph describe the various data rates and SNR requirements of the host of systems. Note that the SNR requirement is defined as the SNR required to maintain a packet error rate (PER) of $10^{-2}$ with a 1,000-byte (8,000 bit) packet; this 1% PER threshold is a stan-

---

[6]A small loss occurs when the approximation $\log_2(1 + x) \approx \log_2(e) \cdot x$ is close; this occurs only for small $x$.

[7]$E_s = P/B_S = P \cdot \rho/B_F$.

**Table 2.11. Wireless Ethernet Spreading Ratios.**

| Scheme | Code Level | Waveform Level (75% excess bandwidth) |
|--------|------------|---------------------------------------|
| Barker-1 | 22 | 40.00 |
| Barker-2 | 11 | 20.00 |
| CCK-5.5 | 1 | 1.82 |
| CCK-11 | 1 | 1.82 |
| PBCC-5.5 | 2 | 3.64 |
| PBCC-11 | 1 | 1.82 |
| PBCC-22 | 1 | 1.82 |

dard measure of "robustness" used by the IEEE 802.11 committee in deliberations leading to the selection of standards.

The graph in Figure 2.19 shows how the superior error control properties of the PBCC method of signal generation can be used to improve robustness (i.e., SNR requirements) or user data rate. It is also interesting to see that the existing IEEE 802.11b standard, which is widely deployed in FCC-certified products, violates the Massey spread spectrum result in terms of Shannon theory. The reason for this dis-

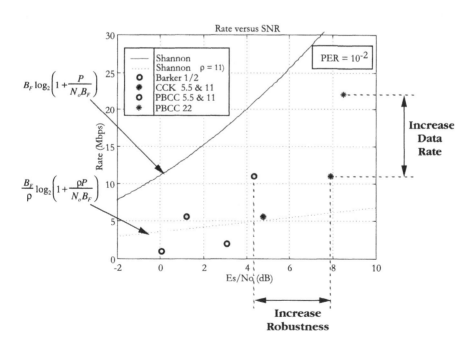

**Figure 2.19.** Performance wireless Ethernet relative to the Shannon limit.

crepancy is explained by the pragmatism of the FCC regulatory body, the FCC's broader definition of spread spectrum, as well as the strictness of Massey's theoretical result. Without such flexibility on the part of the FCC, there would be no high-performance wireless Ethernets.

## 2.6    Range versus Rate Wireless Local Area Networks

To determine the effectiveness of a family of modulation and coding options for wireless Ethernet applications, it is useful to understand how data throughput and distance are traded. In this section, a mathematical model is presented that allows a rational comparison of IEEE 802.11g proposals. In this study the legacy CCK systems are compared to the PBCC, CCK/OFDM, and 11a/OFDM. The comparison demonstrates that whereas PBCC and 11a/OFDM follow similar rate versus range curves in the 2.4-GHz band, the additional overhead required for 802.11b backward compatibility of the CCK/OFDM has a severe rate versus range penalty.

This section is organized in two parts. In the first part, the background information required to compute the range and throughput of a wireless Ethernet system is described. In the second part, a comparison of various alternatives considered by Task Group G is presented. The analysis shows the superiority of the PBCC-based systems over the CCK/OFDM systems.

The IEEE 802.11a standard is defined only for the 5-GHz band, however, Task Group G has been considering a proposal for an "11a-like" scheme known as CCK/OFDM. Thus it is prudent to consider the capabilities of 11a/OFDM if it were transmitted in the 2.4-GHz band. Of course, if the 802.11a signal is transmitted in the 2.4-GHz band it would not be backward compatible with the existing base of IEEE 802.11b networks because the preamble structures of the two signaling sets are incompatible. The introduction of pure 11a/OFDM signals in the 2.4-GHz band could be quite disruptive because the two networks would cause mutual interference if used in an overlapping band of frequencies.

The idea of CCK/OFDM was introduced by Intersil as a method of combining 802.11a signals in a backward compatible way with the single-tone modulations (i.e., Barker, CCK, and PBCC) that are the basis of IEEE 802.11b. The scheme involves transmitting an IEEE 802.11b preamble followed by a transition to the OFDM blocks defined in the IEEE 802.11a standard. Unfortunately, the backward compatible requirement makes the overhead of the CCK/OFDM solution excessive. For the highest mandatory rate, in additive white Gaussian noise (AWGN), PBCC-22 achieves a throughput of 12.8 Mbit/s at a range that is 95% of the CCK-11 system whereas the CCK/OFDM-24 achieves 13.0 Mbit/s (a trival improvement in rate) at a range that is only 76% as far. (The notation XXX-NN denotes an "XXX" signal with a maximum instantaneous rate of "NN" Mbit/s.) In terms of area, these factors represent 90% and 58% coverage, respectively. With 100 ns of multipath, the range numbers become 92% and 74%.

It is interesting to note that 11a/OFDM signals, which do not suffer from the large overhead required to be backward compatible with the 11b preamble, have the same ranges as CCK/OFDM in the 2.4-GHz band but much higher throughput. For example, with 11a/OFDM-24 the throughput is 18.5 Mbit/s. The curves for PBCC and an 11a/OFDM system (used in the 2.4-GHz band) shown in Figure 2.23 demonstrate that for ranges up to 60% of the CCK-11 range, the two schemes are very competitive, whereas the CCK/OFDM system significantly lags both solutions in all cases.

Furthermore, IEEE 802.11a systems are designed for the 5-GHz bands; these higher frequencies experience a penalty because of the higher frequency. This factor predicts that IEEE 802.11a systems will have range problems compared with 2.4-GHz systems at the same power levels and throughput.

### 2.6.1    Background Development

The calculation of user data rate or throughput versus distance involves several components that include:

- Calculation of symbol signal-to-noise ratio ($E_s/N_o$) required for maximal operational Packet Error Rate (PER)
- Translation of waveform signal power to symbol energy
- Determination of receiver noise floor power spectral density ($N_o$) and receiver sensitivity
- Formulation of propagation loss model that relates receiver signal power to distance
- Determination of the maximum throughput of the system including effects of preambles and acknowledgments
- Understanding of effects of multipath distortion on receiver performance

*2.6.1.1    Symbol SNR and PER*    In bandpass digital transmission, a basic concept is the discrete time, two-dimensional symbol. In wireless Ethernet applications, for example, phase shift keying (PSK) and quadrature amplitude modulation (QAM) symbols are transmitted by the sender to convey the intended message. At the receiver, a detection process is used to process the corrupted symbol to determine the message that was transmitted. The corruption of the symbol can include both noise and signal distortion. The noise in the receiver is typically a function of how well the receiver radio can amplify the very small receive signal to bring it to a level that is required by the detection process. The bulk of the noise is modeled as additive white Gaussian noise because the source of the noise is wide band (relative to the signal). The dominant form of signal distortion can be attributed to multipath distortion that arises from multiple reflections of the signal during propagation. The symbol signal-to-noise ratio (SNR) relates the average symbol signal power $E_s$ to the variance of the symbol noise $N_o$ (i.e., the noise in 2 dimensions). For a PSK signal, the symbol energy is constant, $E_s = A^2$, where $A$ is the radius of the circle. For QAM, the symbol energy is generally not constant; the average symbol energy $E_s = A^2$ for 4-QAM (which is the same as QPSK) and $E_s = 5A^2$ for 16-QAM. In Figure 2.20, an 8PSK symbol with $E_s/N_o = 10$ dB is shown.

The effect of the $E_s/N_o$ value on system performance is reflected in the PER of the detector. In the IEEE 802.11 working groups, a threshold PER of $10^{-2}$ (1 packet error in 100 packet transmissions) is considered the maximum acceptable value. Note that because of the incorporation of a reliable error detection code within the body of the packet, it can be assumed that an error-corrupted packet will be detected and rejected (and typically retransmitted). When the PER rises above the threshold, the system typically backs down to a more reliable, albeit slower, transmission mode. The PER is also a function of packet length; for small BER (bit error rate) the PER is approximately N*BER where N is the length of the packet in bits. Thus a packet with 1,000 bytes of data and a PER less than $10^{-2}$ requires a BER of less than $1.25 \times 10^{-6}$.

The detector performance is affected by the choice of transmission signal constellation set and the form of forward error control (FEC) designed into the transmission system as well as the detection algorithm used at the receiver. In Table 2.12, the value of $E_s/N_o$ required to achieve a PER of $10^{-2}$ in AWGN is given. For example, the table shows that the CCK-11 system requires at least 7.8 dB of $E_s/N_o$

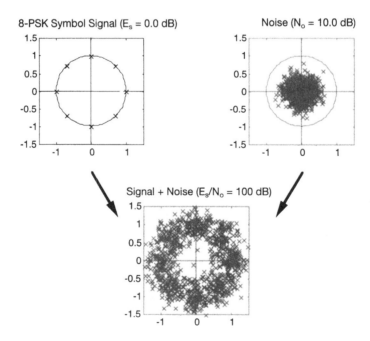

**Figure 2.20.** Signal plus noise in 8PSK.

Table 2.12. Range versus Rate Data, AWGN

| | Mod | Max Rate | Max Throughput* | $E_s/N_o$ (PER:10e-2) | $E_o/N_o$ (PER: 10e-2) | $E_o/E_s$ | Range ($\nu = 3.3$) |
|---|---|---|---|---|---|---|---|
| Item | Mbps | Mbps | dB | dB | dB | | |
| 1 | CCK-5.5 | 5.50 | 4.7 | 4.8 | 4.8 | 0.0 | 123 |
| 2 | CCK-11 | 11.00 | 8.1 | 7.8 | 7.8 | 0.0 | 100** |
| 3 | Uncoded QPSK | 22.00 | 12.8 | 13.5 | 13.5 | 0.0 | 67 |
| 4 | PBCC-5.5 | 5.50 | 4.7 | 1.3 | 1.3 | 0.0 | 157 |
| 5 | PBCC-8.25 | 8.25 | 6.3 | 1.3 | 3.1 | 1.8 | 142 |
| 6 | PBCC-11 | 11.00 | 8.1 | 4.3 | 4.3 | 0.0 | 128 |
| 7 | PBCC-16.5 | 16.50 | 10.7 | 4.3 | 6.1 | 1.8 | 113 |
| 8 | PBCC-22 | 22.00 | 12.8 | 8.5 | 8.5 | 0.0 | 95 |
| 9 | PBCC-33 | 33.00 | 15.9 | 8.4 | 10.2 | 1.8 | 85 |
| 10 | PBCC-49.5 | 49.50 | 18.9 | 11.4 | 13.2 | 1.8 | 69 |
| 11 | PBCC-66 | 66.00 | 20.9 | 14.4 | 16.2 | 1.8 | 56 |
| 12 | CCK/OFDM-6 | 6.00 | 5.0 | 1.2 | 2.9 | 1.7 | 141 |
| 13 | CCK/OFDM-12 | 12.00 | 8.4 | 4.3 | 6.0 | 1.7 | 113 |
| 14 | CCK/OFDM-24 | 24.00 | 13.0 | 10.0 | 11.7 | 1.7 | 76 |
| 15 | CCK/OFDM-36 | 36.00 | 15.9 | 13.2 | 14.9 | 1.7 | 61 |
| 16 | CCK/OFDM-48 | 48.00 | 17.9 | 17.6 | 19.3 | 1.7 | 45 |
| 17 | CCK/OFDM-54 | 54.00 | 18.5 | 18.9 | 20.6 | 1.7 | 41 |
| 18 | 11a/OFDM-6 | 6.00 | 5.6 | 1.2 | 2.9 | 1.7 | 141 |
| 19 | 11a/OFDM-12 | 12.00 | 10.5 | 4.3 | 6.0 | 1.7 | 113 |
| 20 | 11a/OFDM-24 | 24.00 | 18.6 | 10.0 | 11.7 | 1.7 | 76 |
| 21 | 11a/OFDM-36 | 36.00 | 25.2 | 13.2 | 14.9 | 1.7 | 61 |
| 22 | 11a/OFDM-48 | 48.00 | 30.5 | 17.6 | 19.3 | 1.7 | 45 |
| 23 | 11a/OFDM-54 | 54.00 | 32.5 | 18.9 | 20.6 | 1.7 | 41 |

*1,000-Byte Packets with Preamble, 1 SIFS, 1 CCK-11 ACK with Preamble, 1 DIFS
**Reference range = 100

for an acceptable PER whereas the PBCC-11 system requires 4.3 dB of SNR. This 3.5-dB improvement in SNR is a direct consequence of the 64-state Binary Convolutional Code (BCC) [8] specified in the IEEE 802.11b standard for PBCC transmission [1]. Note that the OFDM-12 system, which incorporate a similar 64-state code, has the same coding gain advantage over CCK-11. All of these systems use a QPSK signal set and transmit at a rate of 1 bit-per-symbol because of the presence of a rate 1/2 FEC encoder.

The higher-rate systems incorporate various signal sets and FEC codes. Consider the systems that transmit 2 bits-per-symbol. As a reference, uncoded QPSK requires a threshold $E_s/N_o$ of 13.5 dB. The PBCC-22 system combines 8PSK modulation with a 256-state BCC with a 2/3 code rate. The threshold for PBCC-22 is 8.5 dB, an improvement of 5 dB over uncoded QPSK; this 5-dB improvement is known

**Table 2.13. Range versus Rate Data, AWGN plus Multipath Distortion (100 ns)**

| Item | Mod Mbps | Max Rate Mbps | Max Throughput* dB | $E_s/N_o$ (PER:10e-2) dB | $E_o/N_o$ (PER: 10e-2) dB | $E_o/E_s$ | Range ($\nu = 3.3$) |
|---|---|---|---|---|---|---|---|
| 1 | CCK-11 | 11.00 | 8.1 | 11.1 | 11.1 | 0.0 | 100** |
| 2 | PBCC-11 | 11.00 | 8.1 | 7.0 | 7.0 | 0.0 | 133 |
| 3 | PBCC-22 | 22.00 | 12.8 | 12.3 | 12.3 | 0.0 | 92 |
| 4 | CCK/OFDM-12 | 12.00 | 8.4 | 8.2 | 9.9 | 1.7 | 109 |
| 5 | CCK/OFDM-24 | 24.00 | 13.0 | 13.8 | 15.5 | 1.7 | 74 |
| 6 | 11a/OFDM-12 | 12.00 | 10.5 | 8.2 | 9.9 | 1.7 | 109 |
| 7 | 11a/OFDM-24 | 24.00 | 18.6 | 13.8 | 15.5 | 1.7 | 74 |

*1,000-Byte Packets with Preamble, 1 SIFS, 1 CCK-11 ACK with Preamble, 1 DIFS
**Reference range = 100

as the coding gain. The OFDM-24 systems use 16-QAM symbols with the same 64-state BCC as OFDM-12; the threshold $E_s/N_o$ is 10.0 dB, a 3.5-dB coding gain over uncoded QPSK.

*2.6.1.2  Signal Power to Symbol Energy, Receiver Noise, and Sensitivity*    The signal and noise energy collected at the radio and baseband processor is a function of several factors. With the proper design of transmit signal and receiver structures, incorporating such concepts as "matched filtering," the symbol signal-to-noise ratio will satisfy the equation

$$E_s/N_o = \frac{P_R T_s}{N_o}$$

where $P_R$ is the receive signal waveform power, $T_s$ is the symbol period, and $N_o$ is the noise floor power spectral level.

Intuitively, the symbol energy is derived from the product of signal power (energy per second) and a symbol period (seconds). Notice that such factors as "excess bandwidth," which are important in system design, do not play a role in the equation that matches signal power to symbol energy.

The noise level $N_o$ of the receiver is difficult to estimate analytically because many factors are needed. Such factors include the "noise figure" of the receiver amplifiers and other physical quantities. The fact that the noise floor level (i.e., the power spectral density height) and the symbol noise variance (i.e., the 2-dimensional noise variance) are the same is due to the fact that white noise has the interesting property that the amount of noise is "the same in all directions." If white noise with a power spectral density level of $N_o$ is passed through a filter with impulse response $h(t)$ or transfer function $H(f)$, then the output power is equal to where

$$\|h\|^2 = \int_{-\infty}^{\infty} |h(t)|^2 \, dt = \int_{-\infty}^{\infty} |H(f)|^2 \, df$$

independent of the shape of $h(t)$ or $H(f)$. (In fact, one could use this as a definition of "white" noise.) For this reason, and other related calibration problems, rather than attempt to find an absolute value for the noise floor and the range, we prefer a relative analysis. This approach allows us to compare systems with respect to a known solution, the basic IEEE 802.11b, 11 MHz system.

In our analysis, we take CCK-11 as the base system that is used to set a "stake in the ground" from which other systems are compared. We define a new quantity $E_o$ that will account for factors such as symbol rate and power overhead. The CCK-11 system has a symbol period $T_s = 91$ ns (i.e., the symbol frequency is 11 MHz). When the various systems are compared in terms of range, the ratio of the symbol period to the period of CCK must be considered; for CCK-11, we take $E_o = E_s$ or $E_o/E_s = 1$ (= 0 dB). For PBCC-22, which uses the same symbol rate, $E_o/E_s = 1$ (= 0 dB) also. However, other PBCC modes, such as PBCC-33, use a faster symbol rate of 16.5 MHz, $T_s = 61$ ns, to increase the data rate. In these modes, the bandwidth is preserved by decreasing the excess bandwidth to about 20% from the 80% of typical CCK-11 and PBCC-11 systems. In this case, the nontrivial ratio of symbol periods makes $E_o/E_s = 3/2$ (= 1.76 dB).

In the case of OFDM systems the equivalent symbol period is based on a 12 MHz, $T_s = 83$ ns period. This accounts for a factor of 12/11 (= 0.38 dB) in the calculation of $E_o$. The reasoning for the 12-MHz value can be seen in many ways. For example, the OFDM systems use 48 tones to convey data. Each of the tones is allocated an equal fraction of the transmit power (ideally each tone would receive 1/48 of the power; in fact, each tone gets 1/52 of the power; more on this later) and uses a long symbol period. The symbol period for each tone is 4 μs. This period is obtained via a 64-point FFT that is cyclically extended by 25% (16 terms) to 80 points and clocked with a 20-MHz clock, resulting in a 250-kHz symbol frequency. The 12 MHz follows from the fact that 48 independent tones generating 250k symbols per second will generate 12M symbols per second in total.

There is another factor that must be considered in the calculation of $E_o$ for OFDM systems. This factor is the OFDM signal power overhead, which results from two sources. The first source is the fact that 52 equal power tones are transmitted, but 4 of the tones are used for modem tracking functions and do not carry user information; this results in a factor of 52/48 (= 0.348 dB). The other source is a consequence of the cyclic extension technique for mitigating the effects of multipath to minimize the occurrence of intersymbol interference (ISI). The transmitted tones are orthogonal (the "O" in OFDM) over the 64 points (not the 80) or 3.2 μs (not the full symbol period of 4 μs). The receiver uses this subinterval of 3.2 μs in the detection process and thus sacrifices 5/4 (= 0.969 dB) of the received signal power.

Thus, for OFDM systems, the calculation of $E_o/E_s = 65/44$ (= 1.695 dB); this includes both the symbol rate difference and the signal power overhead.

*2.6.1.3 Propagation Loss*    The signal power observed at the input to the receiver radio is a function of several factors including transmit signal power, antenna gain, and propagation loss from the channel. A common model for propagation loss as a function of distance $d$ takes the form

$$L(d) = c \cdot d^v$$

where the exponent $v$ is the critical parameter of the loss model. In free space, with a spherical radiation of transmit power, the exponent $v = 2$ because the area of the surface of a sphere grows with the square of the radius. In less ideal situations, such as in a building with walls and such, a larger value for the exponent $v$ would be observed. In the IEEE 802.15 committee, a model for propagation loss in Bluetooth systems assumes a free space model up to 8 m and a $v = 3.3$ exponent for larger distances

$$L(d) = \begin{cases} \left(\dfrac{4d_1\pi}{\lambda}\right)^2 \left(\dfrac{d}{d_1}\right)^2, & d \le d_1, \\[4mm] \left(\dfrac{4d_1\pi}{\lambda}\right)^2 \left(\dfrac{d}{d_1}\right)^v, & d \ge d_1, \end{cases} \tag{2.10}$$

where the crossover distance is taken to be $d_1 = 8$ and where the wavelength at 2.4 GHz is $\lambda = 0.1224$ meters. Note that the loss function is continuous in the distance parameter $d$ [13]. This model is derived in Ref. 14, and supported by Eq. 3.1 [page 71, Ref. 15]; it has been adopted by the IEEE 802.11 committee as well.

In this section, the IEEE 802.15/802.11 model at large distance is assumed, i.e., $v = 3.3$. To normalize relative to CCK-11, the waveform signal-to-noise ratio

$$P_R/N_o = \frac{c_o}{d^{3.3}}$$

where the constant

$$c_o = \frac{d_o^{3.3}(E_s/N_o)}{T_s}$$

is determined by setting $d_o = 100$, $E_s/N_o$ is equal to the SNR for CCK-11 that has a PER of $10^{-2}$ (i.e., 7.8 dB) and $T_s = 91$ ns.

Note that choosing $d_o = 100$ forces the range of CCK-11 to be the normalized range of 100. This can be used to estimate the range of other systems once the absolute range of CCK-11 is known. For example, if a realized system has a CCK-11 range of 40 m, then the absolute range for other systems such as PBCC-11 can be estimated. In this case, Table 2.12 indicates a normalized range of 128 (i.e., 28% more) for PBCC-11; this translates into an absolute range of 51.2 m. Similarly, a

PBCC-22 system will reach 38 m, an X/OFDM-12 system will have a range of 45.2 m, and X/OFDM-24 will have 30.4 m reach. With a multipath distortion of 100 ns, the rate versus range performance is shown in Table 2-13.

***2.6.1.4  Rate and Throughput***   It is well known that, in packet systems such as IEEE 802.3 and 802.11, the user data rate is smaller than the maximum instantaneous data rate of the transmission system. In the IEEE 802.11 Medium Access Control (MAC) protocol, a successful data packet transmission is followed by an acknowledgment packet. This overhead is in addition to the other factors such as guard intervals (so-called SIFS and DIFS) and packet preambles and postambles. For reasons of clarity, it is assumed that the acknowledge packets are fixed length at all rates according to Table 2.14.

The throughput of a system is a function of the transmission system, instantaneous rate, and packet length. In this section, packets are assumed to be long, 1,000 bytes in length; this is an optimistic assumption. In addition, this analysis does not account for other forms of MAC overhead such as the MAC header, data error detection, and security such as required for WEP.

In Table 2.12, the throughput for the various choices are listed. As an example of the calculation, consider the transmission of 1,000 bytes (8,000 bits) of data with CCK-11 or PBCC-11. The total transmission time will be $T_{total} = 262 + 8000/11 = 989.27$ μs, yielding a throughput of $R = 8,000/T_{total} = 8.0867$ Mbit/s.

### 2.6.2  Calculation of Rate Versus Range

***2.6.2.1  Rate and Range Data***   The signal-to-noise ratio calculation can be summarized by the equations that relate transmit power to receive power

$$P_R = \frac{PE_oT}{L(d)}$$

and symbol energy to receive power

$$\frac{E_s}{N_o} = \frac{P_R\delta E_oPT_s\delta E_oT}{N_o}$$

**Table 2.14.  Packet Overhead**

| Mod | Preamble | Postamble | DIF | ACK* | Total |
|---|---|---|---|---|---|
| | μs | μs | μs | μs | μs |
| CCK & PBCC | 96 | 0 | 50 | 116 | 262 |
| CCK/OFDM | 108 | 6 | 50 | 116 | 280 |
| 11a/OFDM | 20 | 0 | 34 | 40 | 94 |
| *ACK: Preamble, Data, SIFS | | | | | |

**Table 2.15.** $E_o$ to $E_s$ Translation

| Mod | Rates | $\delta P$ | $\delta T$ |
|---|---|---|---|
| 1 CCK | all | 1 (0 dB) | 1 (0 dB) |
| 2 PBCC | (5, 5, 11, 22) | 1 (0 dB) | 1 (0 dB) |
| 3 PBCC | (8, 25, 16.6, 33, 49.5, 66) | 1 (0 dB) | 22/33 (–1.76 dB) |
| 4 OFDM | all | 48/65 (–1.32 dB) | 11/12 (–.36 dB) |

where $\delta E_o P$ reflects the power overhead and $\delta E_o T$ accounts for symbol clock change relative to the reference (in this section, 11 MHz for CCK-11). For the various systems, Table 2.15 gives the power factors that are the basis of the equation

$$E_s = E_o \delta_P \delta_T, \ E_o = P_R T_s$$

The power overhead $\delta_P \leq 1$, always bounded by 1, has the effect of reducing the symbol energy available for detection from the power received by the radio.

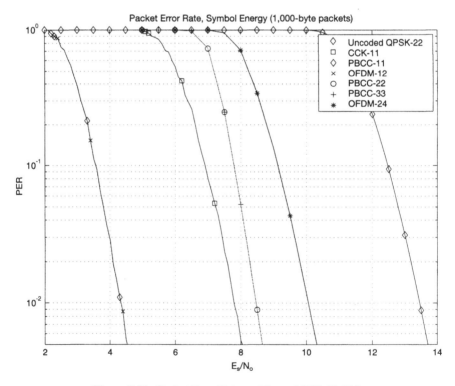

**Figure 2.21.** Packet Error Rate vs. Channel SNR ($E_s/N_o$).

The symbol clock parameter $\delta_T$ is the ratio of the symbol periods (or symbol frequencies) relative to the base, in this case 11 MHz (i.e., the symbol rate of CCK-11). In this section, $\delta_T \leq 1$ because the symbol rates considered are 11, 12, and 16.5 MHz. In Figure 2.21, selected $E_s/N_o$ curves are displayed. These curves show that with this notion of SNR, the PBCC-11 and OFDM-12 systems follow the same curve and have a significant coding gain, about 3.5 dB at a PER of $10^{-2}$ compared with CCK-11. Similarly, the PBCC-22 and PBCC-33 curves are identical on this graph, requiring a fraction of a dB of additional SNR when compared to CCK-11. When one accounts for power overhead and clocking rate differences, one obtains the graph shown in Figure 2.22. On this scale, PBCC-33 moves 1.76 dB to the right because of the higher symbol clock frequency and OFDM-12 and OFDM-24 move 1.70 dB to the right because of the power overhead and clock difference.

The rate and range data for all modes considered in this section are presented in Table 2.13 for AWGN. In Table 2.13, data for channels with 100-ns multipath distortion, generated via the IEEE 802.11 multipath model [9], are presented. This data is displayed in Figures 2.23 and 2.24. In Figure 2.25, the throughput versus area

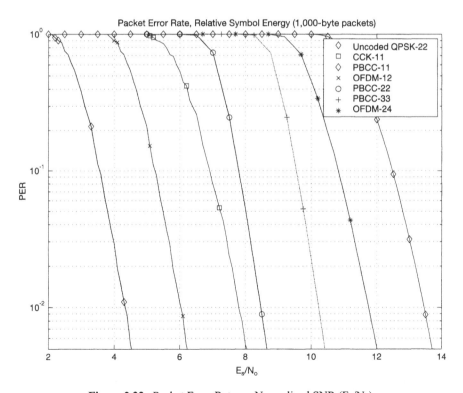

**Figure 2.22.** Packet Error Rate vs. Normalized SNR ($E_o/N_o$).

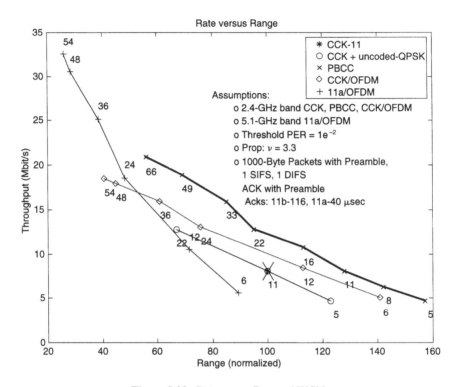

**Figure 2.23.** Rate versus Range, AWGN.

coverage is shown. In Figure 2.26, the throughput versus range in comparison to 5-GHz 802.11a is shown.

These graphs show the superiority of the PBCC-based systems over the CCK/OFDM systems. For the highest mandatory rate, PBCC-22 achieves a throughput of 12.8 Mbit/s at a range that is 95% of the CCK-11 system whereas CCK/OFDM-24 achieves 13.0 Mbit/s at a range that is 76% in AWGN. In terms of area, these factors are 90% and 58% coverage, respectively. With 100 ns of multipath, the range numbers become 92% and 74%.

It is interesting to consider the case of an 11a/OFDM signal if it were used in the 2.4-GHz band; such a hypothetical system would not be backward compatible with 802.11b. However, this modulation would not suffer from the large overhead required to be backward compatible with the 11b preamble; it has the same ranges as CCK/OFDM but much higher throughput. For example, for 11a/OFDM-24, the throughput is 18.6 Mbit/s whereas CCK/OFDM-24 has only a 13.0-Mbit/s throughput. The curves for PBCC and 2.4-GHz 11a/OFDM, shown in Figure 2.23, demonstrate that, for ranges up to 60% of the CCK-11 range, the two schemes are very

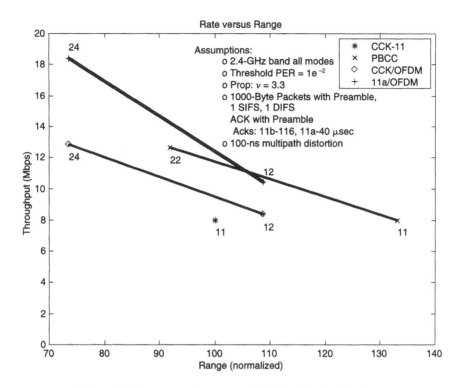

**Figure 2.24.** Rate versus Range, AWGN + Multipath Distortion.

competitive, whereas the CCK/OFDM system significantly lags both solutions in all cases.

Furthermore, IEEE 802.11a systems are designed for the 5.1-GHz (and higher) U-NII bands (these bands include the 2 indoor bands 5.15- to 5.25-GHz band and 5.25- to 5.35-GHz band and an outdoor band 5.725-5.825 GHz in the US). The loss model given in Eq. (2.10) shows a penalty of over 4.3 (= 6.3 dB) in received signal power due to the higher frequency (i.e., shorter wavelength). It is this factor that moves the 11a curves in Figure 2.23 to the left in Figure 2.26. This shows why 5.2-GHz systems have range problems compared with 2.4-GHz systems at the same power levels and throughput.

## 2.7   Conclusions

This chapter considers the history, development, and future of high-speed wireless Ethernet in the 2.4-GHz ISM band. Networks that allow users to connect to net-

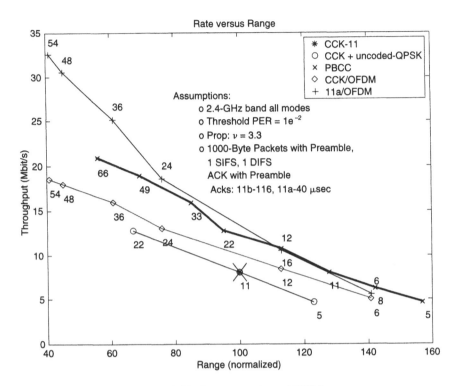

**Figure 2.25.** Rate versus Area, AWGN.

works without wires and with high throughput have recently become popular and show the potential for exponential growth in the coming years.

The birth of wireless Ethernet began over a decade ago with the work of the IEEE 802.11 wireless networking standards body. This group developed the technology behind the very successful IEEE 802.11b standard that has shown explosive growth over the last couple of years.

This chapter considers the origins of the "11b" standard and includes an introduction to the Medium Access Control (MAC) technology including a description of the MAC header structure. The chapter describes the physical layer technology specified in the 11b standard including the CCK and PBCC modes. An extension of the 11b technology developed by Alantro Communications (now a part of Texas Instruments) is described; this extension provides a "double the data rate" (22 Mbit/s) mode that is fully backward compatible with existing 11b networks.

The chapter also discusses the role and limitations of spread spectrum communications in wireless Ethernet.

A comparison in terms of range versus rate is given. The comparison includes Intersil's CCK/OFDM modulation as well as the 802.11a standard in the 5-GHz bands.

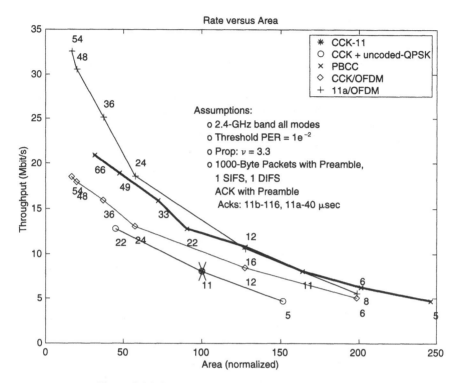

**Figure 2.26.** Rate versus Range, AWGN w/ 5-GHz data.

Currently, Texas Instruments is shipping wireless Ethernet chips that fully implement the 11b standard, with both CCK and PBCC modes, and include the PBCC-22 extension.

## Acknowledgments

The authors thank Dick Allen and Anuj Batra for their very valuable feedback in the preparation of sections of this chapter. They are also grateful to the editor, Benny Bing, for his detailed feedback; his efforts greatly improved this chapter.

## References

1. I. 802.11, "1999 high rate," Tech. Rep. Reference number ISO/IEC 8802-11, IEEE Std 802.11, The Institute of Electrical and Electronic Engineers, 1999. ISO/IEC 8802-11, IEEE Std 802.11 ANSI, Information technology—Telecommunications and information

exchange between systems—Local and metropolitan area networks—Specific requirements—Part 11: Wireless LAN Medium Access Control (MAC) and Physical Layer (PHY) specifications.

2. I. 802.11, "1997 low rate," Tech. Rep., The Institute of Electrical and Electronic Engineers, 1997.

3. C. Heegard, J. S. T. Coffey, S. Gummadi, P. A.Murphy, R. Provencio, E. J. Rossin, S. Schrum, and M. B. Shoemake, "High performance wireless ethernet," *IEEE Communications Magazine*, Nov. 2001.

4. Cahners, "Wireless markets," Tech. Rep., Cahners/In-Stat, 2000.

5. J. G. Proakis, *Digital Communications*. New York: McGraw-Hill, 3rd ed., 1995.

6. C. Heegard and S. Wicker, *Turbo Coding*. No. ISBN: 0-7923-8378-8, Boston: Kluwer Academic, Jan. 1999.

7. D. Slepian, "Group codes for the gaussian channel," *Bell System Technical Journal*, vol. 47, pp. 575–602, Apr. 1968.

8. S. B. Wicker, *Error Control Systems for Digital Communications and Storage*. No. ISBN: 0132008092, Englewood Cliffs, NJ: Prentice Hall, 1995.

9. I. 802.11, "Multipath model for comparison criteria," Tech. Rep. IEEE 802.11-00-282r2, The Institute of Electrical and Electronic Engineers, 2000.

10. J. L. Massey, "Towards an information theory of spread-spectrum systems," in *Code Division Multiple Access Communications* (S. G. Glisic and P. A. Leppanen, eds.), pp. 29–46, Boston: Kluwer, 1995.

11. H. Nyquist, "Certain topics in telegraph transmission theory," *Transactions of the AIEE*, vol. 47, pp. 617–644, 1928.

12. C. E. Shannon, "A mathematical theory of communication," *Bell System Technical Journal*, vol. 27, pp. 379–423, 623–656, Oct. 1948.

13. I. 802.15, "Multipath model for comparison criteria," Tech. Rep. IEEE 802.15-99-138r0, The Institute of Electrical and Electronic Engineers, 1999.

14. W. L. Stutzman and G. A. Thiele, *Antenna Theory and Design*. No. ISBN: 0471025909, New York: John Wiley & Sons, 2nd ed., 1996.

15. T. S. Rappaport, *Wireless Communications*. No. ISBN: 0133755363, New Jersey: Prentice Hall, 1st ed., 1996.

# Chapter 3

# The 5-GHz IEEE 802.11a Wireless LAN

James C. Chen, Ph.D
*Atheros Communications, Inc.*[1]

Wireless local area networks have come a long way from their humble roots. What started out as a way for vertical industries to transmit data in warehouses and factory floors has grown into a cost-effective means for enterprises to network increasingly mobile workers for increased productivity. Last year approximately 7 million wireless LAN units were sold, generating an estimated $1 billion market. Although such results seem impressive, wireless LANs have yet to realize their full potential. Systems built to the IEEE 802.11a standard will soon appear in the market to take advantage of higher data rates and more frequency channels for even greater performance. In this chapter, we present an introduction to 802.11a systems, emphasizing the fundamentals of Orthogonal Frequency Division Multiplexing (OFDM). In addition, measured 5-GHz 802.11a performance data are presented for the first time. The range performance of 5-GHz 802.11a systems is measured in terms of data link rate and throughput. These results are then used to calculate 802.11a system capacity. Here, 802.11a not only provides higher end user speeds but also allows reductions in wireless LAN deployment costs.

## 3.1 Introduction

Many events conspired to produce the success of wireless LANs. The advent of the IEEE 802.11b standard that achieved Ethernet-equivalent speeds, the creation of the Wireless Ethernet Compatibility Alliance (WECA) as an industry forum that pushed for Wi-Fi™ interoperability among equipment vendors, and the decision by major notebook makers to integrate wireless LANs into mobile PCs for the mass market all played pivotal roles. There is no reason why such efforts and trends will not continue for 802.11a and for future versions of wireless LANs. As Table 3.1 shows, regulatory and standards bodies have laid a solid foundation for this future. Certain inherent advantages are evident in terms of more frequency spectrum, higher data rates, and more advanced modulation techniques. As such, the resulting benefits to 802.11a users are likely to be very significant and must be carefully studied and understood. The supported data rates are provided in Appendix A.

---

[1] Atheros™ and The Air is Cleaner at 5 GHz™ are trademarks of Atheros Communications, Inc.

**Table 3.1. Approved IEEE standards (valid only in US per FCC regulations).**

|  | 802.11a | 802.11b | 802.11 |
|---|---|---|---|
| Date Approved | September 1999 | September 1999 | July 1997 |
| Frequency Band | 5.150–5.350 GHz, 5.725–5.825 GHz | 2.4–2.4835 GHz | 2.4–2.4835 GHz |
| Available Bandwidth | 300 MHz | 83.5 MHz | 83.5 MHz |
| Number of Nonoverlapping Channels | 12 (Indoor/Outdoor) | 3 (Indoor/Outdoor) | 3 (Indoor/Outdoor) |
| Data Rate per Channel | 6, 9, 12, 18, 24, 36, 48, 54 Mbit/s | 1, 2, 5.5, 11 Mbit/s | 1, 2 Mbit/s |
| Modulation Type | OFDM | DSSS | FHSS, DSSS |

This chapter details these benefits and illustrates 802.11a's performance gains as well as lower deployment cost advantages. It will begin by exploring how the 5-GHz spectrum provides a cleaner environment for wireless LANs because of regulatory rules designed to mitigate potential interference. Next, a short tutorial on the benefits of OFDM, the multiplexing scheme used in 802.11a, is provided. Two sections then follow to illustrate the performance advantages of 802.11a over 802.11b. Specifically, measured range performance data from a typical office environment shows 802.11a's ability to provide a range similar to 802.11b but at two to five times higher data rates. These measurement results are then used in conjunction with a published IEEE model on system capacity to illustrate why 802.11a can provide more system capacity through the availability of more channels.

## 3.2 FCC's 5-GHz Rules Mean a More Interference-Free Environment

Historically, unlicensed bands are rarely interference free. Competing devices from different industries, or even the same ones, inevitably compete to use the same frequency band in their own way. The cases associated with 2.4 GHz are well publicized [1] and need no additional mentions here. The level of concern is genuine and high because potentially large numbers of consumer electronics and Bluetooth devices are being released into the market and no one can guarantee 2.4-GHz wireless LAN users definite relief and peace of mind.

Part of the interference problem lies in the regulations for transmit power levels in the 2.4-GHz Industrial, Scientific, and Medical (ISM) band. For example, in the 2.4-GHz to 2.4835-GHz band, FCC 15.247 rules [2] state that all Frequency Hopping Spread Spectrum (FHSS) and Direct Sequence Spread Spectrum (DSSS) devices can have a maximum peak output power of 1 W. This effectively means that a narrowband FHSS device can interfere with a wideband DSSS device by hopping anywhere inside the latter's wider channel. The unrestricted and high 1-W power level only makes the situation worse. To put this in context, a Bluetooth device transmits a 1-MHz narrowband channel and hops 1,600 times per second over 79 channels in the 2.4-GHz ISM band. It therefore has a very high likelihood of hopping into a stationary 22-MHz channel that a 2.4-GHz DSSS wireless LAN system uses. A similar situation holds for 2.4-GHz cordless telephones as well.

For the 5-GHz Unlicensed National Information Infrastructure (U-NII) band, FCC regulation 15.407 is more expertly crafted. Table 3.2 shows different output power requirements for different parts of the 5-GHz spectrum. Here, maximum ratings refer to peak output power delivered to the antenna. Equivalent Isotropically Radiated Power (EIRP) measures the power transmitted by a directional antenna in the strongest direction. Clearly, the FCC has set different power limits for different parts of the 5-GHz U-NII band. These limits can play a pivotal part in separating and mitigating potential interference for current and future applications. For example, the upper U-NII band from 5.725 to 5.825 GHz is best suited for outdoor fixed broadband wireless access (FBWA) devices, which typically require much higher power levels to reach longer distances. Wireless LANs are still able to operate in this upper 100-MHz spectrum but are well advised not to because of the presence of potential FBWA interference. The lower 200-MHz band (5.15- to 5.35-GHz band) will serve indoor (as well as outdoor) wireless LANs better. The low power rulings here render FBWA applications ineffective. Similar rules have been adopted in Europe as well.

Along with the power limits given above, the FCC has specified power spectral density limits, which force systems with bandwidths narrower than 802.11a to transmit with less power. Furthermore, all U-NII devices must be high-data rate communications devices [2]. This implies that previously mentioned 2.4-GHz narrowband interferers (i.e., cordless phones, low-rate Bluetooth devices) are not likely to find a home in the 5-GHz band. Overall, FCC 5-GHz power rules help mitigate and limit potential interference to 5-GHz wireless LANs, even in the future. Clearly, the 5-GHz band is currently a very clean band for 802.11a wireless LAN operation.

## 3.3 Benefits of OFDM

As stated in Section 3.2, 802.11a systems use the OFDM scheme. To understand OFDM, it is best to first understand some of the side effects of transmitting radio signals over the air. In an ideal radio transmission, a transmitter sends out a signal that reaches the receiver in a single direct path, without any indirect paths reflecting off walls and other objects. In this way, the received signal is an exact copy of the transmitted one. Unfortunately, this is often not the case. In reality, the radio signal is modified as it passes through the channel (the physical space between the transmitter and the receiver). The transmitted signal can be degraded with a combination of effects: It can be attenuated, reflected, refracted, and diffracted and may produce many replicas of its original self. On top of all of this, the channel also introduces noise into the signal and can cause a shift in the carrier frequency (known as the Doppler effect) if the transmitter or receiver is moving. With all of these simultaneous effects, it is a wonder that radio transmission systems work as well as they do!

**Table 3.2. Power regulations in the 5-GHz band.**

|  | 5.15–5.25 GHz | 5.25–5.35 GHz | 5.470–5.725 GHz | 5.725–5.825 GHz |
|---|---|---|---|---|
| US | 50 mW (Max) 200 mW (EIRP) | 250 mW (Max) 1 W (EIRP) | NA | 1 W (Max) 4 W or 200 W (EIRP) |
| Europe | 200 mW (EIRP) | | 1 W (EIRP) | 25 mW (EIRP) |
| Japan | 200 mW (EIRP) | NA | NA | NA |

### 3.3.1 Multipath—Unwanted Echoes

Severe modifications of the signal can occur when the transmitted signal is reflected from objects such as walls, furniture, and other indoor objects. Under such circumstances, the transmitted signal may not have a single direct path to the receiver. Rather, there can be a number of different paths (or multipaths), each of which has a different distance to travel from the transmitter to the receiver and thus experiences a different delay. As a result, the signal can have multiple "echoes" of itself that arrive at the receiver at different moments in time. Thus, from the receiver's point of view, it receives multiple copies of the same signal with many different signal strengths or powers. Figure 3.1 shows a plot of the average received signal power versus time.

The delay spread, $\tau_{max}$, is defined as the maximum time difference between the arrival of the first and last multipath signal seen by the receiver. The delay spread is a function of the transmission environment. Large delay spreads are usually found to be characteristic of large buildings because the distances between the transmitter and reflectors are greater in such big environments. In addition, delay spreads do not show any significant dependencies on the transmission frequency. Measurements show the same behavior for frequencies ranging from 800 MHz to 6 GHz. Typical delay spreads for indoor transmission vary from 40 to 200 ns whereas outdoor values vary from 1 to 20 µs.

To understand the effects of multipath for high-speed data networking, assume that a radio is transmitting a discrete block of digital information (i.e., a symbol) every $T$ time intervals (Fig. 3.1). Under such conditions, a received symbol can potentially be corrupted by echoes of up to $\tau_{max}/T$ previous symbols. This effect is defined as intersymbol interference (ISI). ISI gets worse (i.e., the $\tau_{max}$-to-$T$ ratio increases) as the transmitted bit rate is increased because of $T$ decreasing. If this ratio becomes too large, correcting for multipath becomes a very complicated problem in the receiver.

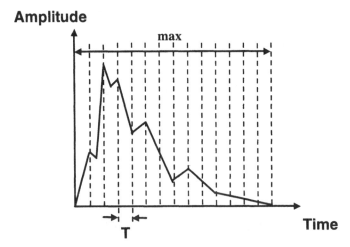

**Figure 3.1.** Multipath can produce signal "echos" that can be delayed by up to $\tau_{max}$.

### 3.3.2 The Multicarrier Approach

One solution to decrease ISI is to somehow decrease $\tau_{max}/T$. Because delay spread is a function of the environment and cannot be altered by the radio, the only recourse is to increase the transmission interval time $T$ between the transmitted symbols. But this slows down the transmission and runs contrary to the mantra of high speed. OFDM finds a way to satisfy both requirements: Instead of transmitting the information using one frequency, or carrier, at an interval of $T$, OFDM divides the transmission among $N$ different subcarriers, each with a transmission interval time lengthened by $N$ (or with $1/N$ as much data). Thus, despite the fact that the data rate for each individual subcarrier has been reduced by a factor of $N$, the parallel transfer of $N$ different transmissions means that the overall transmission rate of the system will remain the same. In addition, $\tau_{max}/T$, with respect to each subcarrier, has been decreased to $\tau_{max}/(T * N)$. This means that each subcarrier is now $N$ times more multipath- and ISI-tolerant. In terms of the IEEE 802.11a standard, $N$ is equal to 52.

Besides increased immunity to ISI, the parallelism introduced by OFDM also has a side benefit of making the transmitted symbol less susceptible to selective frequency fading. Suppose that data is being transmitted using only a single carrier. If the channel introduces interference at this frequency, the entire transmission can fail. This scenario will have a lesser effect on a multicarrier system, because only a few of the subcarriers will typically be affected. Error-correcting codes (requiring redundant bits) can then be used to help restore the actual information lost in these erroneous subcarriers. OFDM systems that employ error-correcting codes to compensate for lost carriers are commonly referred to as Coded OFDM (COFDM).

Despite the added robustness, OFDM's multicarrier approach may not solve all ISI problems. Figure 3.2 illustrates this for one directly received signal (no multipath) and another signal that has been shifted because multipath delay, which is now smaller because of addition of $N$ subcarriers. The overlap of the current symbol with the delayed previous symbol still produces ISI. The remedy for this problem is to provide a guard time at the beginning of each transmitted symbol. This guard time acts as a buffer to allow time for multipath signals from the previous symbol to die away before the information from the current symbol is gathered at the receiver (see Fig. 3.3). The length of the guard interval, therefore, should be equal to at least the delay spread. Typically, most systems set the guard interval to the maximum allowable delay spread plus some margin for error. For example, a maximum indoor delay spread of 200 ns can be more than accommodated by a guard interval of 600–800 ns. Of course, the addition of this guard time is not free because it reduces the available time for symbol transmission.

The most effective guard period to use for OFDM is most commonly referred to as a cyclic prefix (CP). This is nothing more than a direct copy of the end of the symbol placed at the start of the symbol. The extra symbol length now serves as the guard period. A copy is used to preserve the orthogonality of the waveform and prevent intercarrier interference (ICI). Both of these topics are discussed in the next section.

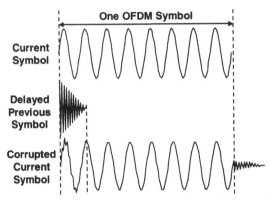

**Figure 3.2.** With no guard time, the delayed transient of the previous symbol can corrupt the current symbol.

### 3.3.3 Where's the Orthogonality?

To begin discussion of the orthogonality component of OFDM, it is best to understand the actual signals and waveforms that are generated and sent by the OFDM transmitter. For each subcarrier, the data to be sent is encoded by a certain modulation scheme. In an OFDM system, the modulation scheme used is usually either Phase Shift Keying (PSK), where the data is represented as different phase offsets of a signal, or Quadrature Amplitude Modulation (QAM), where data is represented by changing both amplitude and phase. Once these sets of amplitudes and phase offsets for each subcarrier are known, they are combined into one composite signal, in the time domain, using an Inverse Fast Fourier Transform (IFFT). The end result of this process is the conversion of individual data bits into a single time domain signal containing a collection of subcarriers. All of these subcarriers together make up one OFDM symbol that is then transmitted into the channel. At the receiver side, the reverse function is performed. Specifically, the separation of subcarrier signals into individual amplitudes and phase offsets is performed with an FFT algorithm.

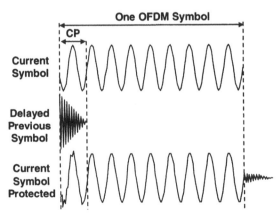

**Figure 3.3.** With a cyclic prefix guard time, the delayed transient of the previous symbol will not corrupt the current symbol.

The orthogonality comes from the precise relationship between the subcarriers that make up one OFDM symbol. Figure 3.4 shows an example of three subcarriers within one OFDM symbol. In this case, all subcarriers have the same phase and amplitude to simplify the illustration. Note how:

- Each subcarrier is exactly an integer number of cycles in a given $T$ time interval. In other words, each subcarrier frequency is an integral multiple of a base frequency (i.e., $f_1 = f_0, f_2 = 2f_0, f_3 = 3f_0$, etc.);
- The number of cycles in a symbol period between two adjacent subcarriers differs exactly by one.

Although simplified, these properties account for the orthogonality between the subcarriers and allow each received subcarrier to have its data bits demodulated independently and free from any other subcarrier interference that may be present.

Another popular way to view this orthogonality is seen in Figure 3.5. This is an equivalent representation of Figure 3.4 in the frequency domain. Each subcarrier's frequency spectrum is represented by a sinc function. One property of this sinc function is that the peak at its center frequency corresponds to the zero at all integer multiples of this frequency. The OFDM receiver can effectively demodulate each subcarrier because, at the peaks of each of these sinc functions, the contributions from other subcarrier sinc functions are zero. Note how all of the subcarriers are packed very tightly against one another (i.e., they do not have extra buffers between them). This allows very efficient use of the frequency spectrum.

At this point, OFDM orthogonality is introduced without any multipath effects. Will its presence derail OFDM's orthogonality qualities? The answer is no, as long as guard time is implemented with CP and the multipath delay is within the guard time. Figure 3.6 illustrates this. A delayed subcarrier with CP still satisfies the orthogonality condition previously mentioned. However, the delayed subcarrier without CP does not. This latter condition leads to a phenomenon known as intercarrier interference (ICI). This means that as each received subcarrier is being demodulated, it will encounter some interference from another carrier because, within the FFT interval, there is no integral number of cycles between the two subcarriers. It is therefore evident that for orthogonality to be preserved, the OFDM subcarriers must have guard times with CP and their precise interrelationships must be carefully controlled or synchronized.

- - - - - Subcarrier 1

———— Subcarrier 2

▬▬▬▬ Subcarrier 3

**Figure 3.4.** An example of 3 subcarriers transmitted in 1 OFDM symbol. For simplicity, the amplitudes and phase offsets of the subcarriers are shown to be the same.

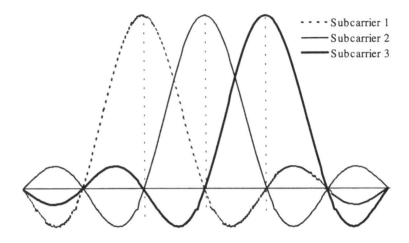

**Figure 3.5.** Orthogonality is achieved when the peak of each subcarrier spectrum occurs at the nulls (zeroes) of all other subcarriers.

### 3.3.4 On Your Mark, Get Set, Synchronize

In an OFDM system, the subcarriers are perfectly orthogonal if they all have an integral number of cycles within the FFT interval. If there is a frequency offset, then the number of cycles in the FFT interval will no longer be an integer and will thus result in ICI. Frequency offsets occur because the signal is usually not transmitted at low baseband frequencies. Rather, it is modulated with a higher radio frequency (e.g., at 5 GHz in IEEE 802.11a). Manufacturing tolerances of the components that make up the transmitter and receiver are usually large enough to incur a significant frequency error between the transmitter and receiver. In many cases this deviation is too large for reliable data transmission and must, therefore, be estimated and compensated. For a 5-GHz OFDM system with a subcarrier spacing of 300 kHz and a negligible degradation of 0.1 dB, the maximum tolerable frequency offset is less than 1% of the subcarrier spacing. This means that the oscillator frequency needs to be about 3 kHz or 0.6 parts per million (ppm) of the 5-GHz carrier. Most oscillators will not be able to meet this requirement. Therefore, frequency synchronization must be applied before the FFT.

Similarly, timing synchronization is also needed. Before the OFDM receiver can demodulate the subcarriers, it has to find out where the symbol boundaries are. This information is used to synchronize the receiver and transmitter time scales. This is essential for the proper removal of the CP and to ensure proper duration of the FFT interval.

Another related timing synchronization issue involves the need to perform sampling-clock synchronization. At the transmitter, the signal produced by the IFFT will be converted into an analog signal with a certain time interval (sampling interval) between each value. After the signal is received, it is down converted. This signal is then sampled to obtain a discrete time signal for subsequent digital processing. These sampling times in the receiver must match very closely with those of the transmitter to avoid performance degradation.

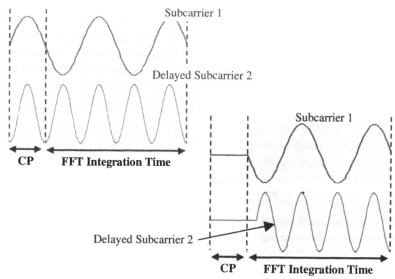

**Figure 3.6.** With no signal in the CP (lower right plot), an integral number of waveform cycles does not exist. Orthogonality is not preserved, and ICI will occur.

To perform frequency and timing synchronization, OFDM systems use special training symbols that are sent from the transmitter and compared with their known values at the receiver. Depending on the outputs of this comparison, the desired timing and frequency offset information is obtained.

## 3.4 802.11a Range Performance

Many studies involving range performance of 802.11a have used theoretical link models and radio wave propagation characteristics to support their claims. Although a theoretical model may offer predictive capability, it does not offer real-world validation. This is especially true in environments in which signal impairments due to multipath are present. Furthermore, all models depend on theoretical assumptions rather than actual radio implementations. This section presents, for the first time, measured 5-GHz 802.11a range performance data collected in a typical office environment. Details of the measurement setup and resulting measurement results are explained below. Identical tests were repeated for a popular 802.11b product as well. The results indicate that 802.11a systems have a range similar to that of 802.11b systems in a typical office environment but with two to five times higher throughput performance.

### 3.4.1 Measurement Setup

The measurement environment was Atheros' Sunnyvale office in California. This is a 265 ft by 115 ft rectangular facility with conference rooms, closed offices, and walls as well as semiopen cubicle spaces. For the 802.11a system, data were sent between two Atheros 802.11a PC card reference designs. One card served as the fixed Access Point (AP), and the other served as a mobile station. Distances of up to 225 ft were measured. The 802.11a reference design employed the Atheros AR5000 chipset and had an output power of 14 dBm to the antenna for both the AP and the mobile station. The PC card reference design used for the 802.11a AP included a reference design external antenna with an average gain of 4 dBi. The 802.11b system under test comprised an AP and PC card from a leading 802.11b manufacturer. This system had an output power of 15 dBm to the antenna for both the AP and the mobile station.

For both systems, the mobile station was moved to the same 80 random locations, which included open cubicle areas, various closed offices, and conference rooms. At each location, the placement of the laptop followed a random orientation to be representative of actual use (i.e., users do not manually adjust the orientation of their mobile stations). A random orientation also lessened the advantages for any antenna gain with respect to a particular orientation. The same orientation at each location was used for both 802.11a and 802.11b systems to maintain a uniform comparison between the two systems.

The range test relied on an optimal rate adaptation method. At each location, 100 broadcast (i.e., nonacknowledged) packets at each data link rate were sent from the AP to the mobile station. This was done to obtain statistically meaningful results. The packet size was fixed at 1,500 bytes, and no fragmentation was used. The mobile station then recorded how many of these packets were received successfully to compute a Packet Error Rate (PER). This measurement technique allowed close monitoring of the physical link performance of both systems without being subject to performance effects due to variability in software (i.e., rate adaptation) or higher layer protocols and applications (e.g., FTP file transfer using TCP/IP). Again, the same measurement method was used for both 802.11a and 802.11b systems.

After all PER measurements were taken, an optimal rate adaptation algorithm was used to determine the data link rate and throughput performance. This was applied to both 802.11a and 802.11b systems. Recall that at each of the 80 measurement locations, packets were sent at all data link rates (i.e., 6, 9, 12, 18, 24, 36, 48, and 54 Mbit/s for 802.11a and 1, 2, 5.5, and 11 Mbit/s for 802.11b) and the associated PERs were recorded. These PERs were then used to compute an effective MAC throughput for each of the data link rates. This calculation accounted for the MAC and physical (PHY) layer overhead and effect of packet retries. The calculation was based on 802.11 specifications for interframe spacings, slot times, PHY layer overhead, etc. An example of this calculation can be found in Appendix B and is summarized in Appendix C. The best throughput was selected for each location, and this process was repeated for all 80 locations.

### 3.4.2 Data Link Rate Results

For each location, the optimal data link rate is defined as the link rate yielding the highest throughput. This determination was repeated for all 80 locations. This process was applied to both 802.11a and 802.11b systems to remove the effect of different software rate adaptation algorithms. For the data link rate measurement, a median filter was applied to the data from each of the 80 locations to smooth the data. The purpose was to produce results that provided a fair representation of the overall range performance. The use of the median filter means that at each link rate, there is an equal number of measured ranges that are less than as well as greater than the median values. The 802.11a and 802.11b median range performances are plotted in Figure 3.7. The data is in "steps" corresponding to the defined data link rates of 54, 48, 36, 24, 18, 12, 9, and 6 Mbit/s.

Two main conclusions can be readily drawn from Figure 3.7:

- 802.11a has similar range compared to 802.11b up to 225 ft in a typical office environment;
- For all distances up to 225 ft in a typical office environment, the data link rates of 802.11a are two to five times better than 802.11b.

Other notable observations are that at the maximum measured distance of 225 ft, 802.11a yielded a 6 Mbit/s rate versus 2 Mbit/s for 802.11b. At the highest 802.11b 11 Mbit/s range (i.e., 107.5 feet), 802.11a still operated at a higher data link rate of 18 Mbit/s. Of course, at closer distances this improvement becomes larger. In actual use, many enterprises are deploying smaller cells with 65 ft radii. This is done for each AP to serve a smaller number of users, thereby providing each user a higher speed. At 65 ft, Figure 3.7 shows that 802.11a improves its speed advantage by delivering a 36 Mbit/s data link rate.

### 3.4.3 Throughput Results

Data link rates provide an insight into how wireless LAN systems trade performance for range. However, another important metric is throughput versus range. Throughput is the actual rate of information that can be transmitted accounting for various overheads. Throughput is dependent on several factors: data link rate, MAC efficiency, measured PER, and packet size. Other factors such as efficiency of higher layer protocols (e.g., TCP/IP), collisions, and the number of users can also affect throughput but were not considered in this analysis.

As described in Section 3.4.1, throughput performance was determined by selecting the best throughput at each location. This process was repeated for all 80 locations. The resulting set of 80 throughput data points was then averaged to smooth the data. This methodology was repeated for both 802.11a and 802.1b systems under test and is plotted in Figure 3.8. The data is not shown to be in "steps" (as compared to Fig. 3.7) because throughput at each defined data link rate can vary depending on the measured PER.

Two main conclusions can be readily drawn from Figure 3.8:

- 802.11a has higher throughput than 802.11b up to 225 ft in a typical office environment;
- For all distances up to 225 ft in a typical office environment, the throughput of 802.11a systems are 2–4.5 times better than 802.11b.

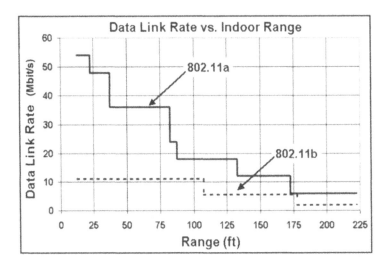

**Figure 3.7** Measured median range performance data for 1,500-byte data packets indicates that the range of 802.11a is similar to 802.11b up to 225 ft in a typical office environment. At 225 ft, 802.11a systems were measured at 6 Mbit/s whereas 802.11b systems were at 2 Mbit/s.

Specifically, at the maximum measured distance of 225 ft, 802.11a yielded a 5.2 Mbit/s rate versus 1.6 Mbit/s for 802.11b. At more realistic deployment distances of 65 ft, 802.11a extends its speed to 21 Mbit/s versus 5.1 Mbit/s for 802.11b. These throughput results are used in the calculations on system capacity in the subsequent sections.

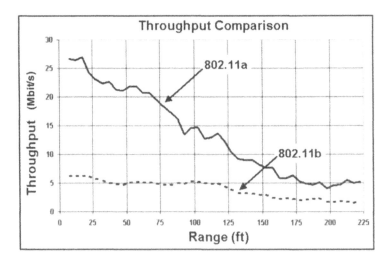

**Figure 3.8** Averaged throughput performance for 1,500-byte data packets. The results indicate that 802.11a throughputs are always at least a factor of 2 times and up to 4.5 times larger than 802.11b systems up to 225 ft.

## 3.5 802.11a System Capacity Benefits

So far the discussion has been limited to measured performance between two nodes, one AP and a mobile station. In a real-world wireless LAN deployment, there are many APs, each simultaneously serving many stations within a given area or cell. A more meaningful question that may be asked is, "Given a deployment of multiple APs, how much throughput does each user receive?" To answer this question, we need to introduce and discuss the issue of system capacity. System capacity refers to the throughput of an entire wireless LAN system comprised of many radio cells (coverage areas). Before we can begin a discussion on 802.11a system capacity versus that of 802.11b, we first need to understand the throughput of a single-cell wireless LAN network.

### 3.5.1 Single-Cell Throughput

For a single mobile station within a cell, the cell throughput is equivalent to the throughput received by the station. For multiple stations in a cell, the average cell throughput is divided equally among the stations (assuming equal sharing among all stations). On the basis of the measured results of Figure 3.8, throughput of the cell is the highest when the mobile station is closest to the center of the cell, or AP, and lowest when it is farthest away. Between these extremes is an average throughput for the entire cell. This average cell throughput represents an average value that the cell can provide to a mobile station irrespective of its location within the cell. On the basis of the measured results of Section 3.4, the average cell throughput of an 802.11a cell with a 225-ft radius in a typical office environment is 9.41 Mbit/s. This is a threefold increase over the throughput of an 802.11b system (3.13 Mbit/s) in the same office environment. For a more realistic cell radius of 65 ft (or a cell size of 130 ft), 802.11a average cell throughput is 4.5 times that of 802.11b—22.6 Mbit/s versus 5.1 Mbit/s. In other words, for an 802.11b system to provide the same total throughput as an 802.11a system, more than four 802.11b APs will have to be deployed (each operating on a unique frequency) in the same area (see Fig. 3.9).

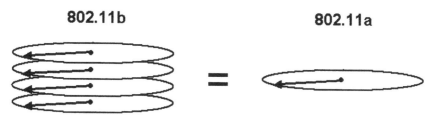

**Figure 3.9.** For a cell radius of 65 ft, more than four 802.11b cells will have to be overlaid on top of each other to achieve the same average throughput of a single 802.11a cell. This assumes that each 802.11b AP can operate on a unique frequency. In reality, this can never be accomplished because 802.11b systems can only operate on 3 distinct channels as mandated by the FCC regulations for the 2.4-GHz unlicensed band.

### 3.5.2 Impact of Co-Channel Interference

Unfortunately, the deployment scenario described in Section 3.5.1 is not possible for 802.11b systems. The reason is that the fourth 802.11b cell will have to operate using one of the previous three channels. The sharing of the same channel between two adjacent cells reduces their average throughput. This effect is referred to as Co-Channel Interference (CCI). Conceptually, it is easy to understand that the key factor in eliminating or reducing CCI is to increase the number of available channels. Figure 3.10 illustrates this point for an eight-cell system deployed using 802.11a and 802.11b. The eight indoor wireless LAN channels allotted for 802.11a by FCC regulations prevent any CCI in this eight-cell system. This is not the case for the eight-cell 802.11b system. Each channel has at least one additional CCI cell for an average of 1.67 CCI cells over all three frequencies.

### 3.5.3 System Capacity Under CCI

One way to evaluate the impact of CCI on average cell throughput is to use a model for system capacity. A system capacity model proposed in 1998 by NEC [3] to the IEEE wireless LAN standardization group was used. Measured range performance data from Section 3.2.2 were included in this model to make its results more indicative of actual 802.11a and 802.11b range performance. There are two mechanisms that model the effect of CCI on system capacity. The Clear Channel Assessment (CCA) mechanism describes the decrease in throughput that results when the AP inside a particular cell has to wait until the channel is available for transmission. The second, "hidden cell" mechanism models how transmissions from undetected cells can corrupt transmission, thereby lowering throughput.

The system capacity model and measured range data were used to evaluate the system capacity for an eight-cell wireless LAN system depicted in Figure 3.10. An eight-cell system was chosen because it is representative of deployments in small and medium-sized enterprises (SMEs). Figure 3.11 shows the result of this analysis. For a typical cell radius of 65 ft (cell diameter of 130 ft), an 802.11a system provides over eight times the average cell throughput (and therefore, eight times the system capacity) of an equivalent 802.11b system. For a given number of users, each user on an 802.11a network will experience eight times the throughput of a user on an 802.11b network. This increase results from the fact that there was no CCI in the 802.11a system because of the availability of eight channels (802.11a systems have eight indoor channels vs. three for 802.11b according to FCC regulations).

The advantages of having more channels carry over to larger systems as well. In multiple-cell systems, such as those shown in Figure 3.12, 802.11a will have CCI cells but a fewer number than 802.11b. If we use channel 1 as the point of reference, the number of CCI cells at one "cell distance" (or the first ring away from the center cell) is 0 for both 802.11a and 802.11b systems. If we extend this distance to the second ring, 802.11a continues to have no CCI cells, whereas 802.11b has six. For the third ring, 802.11a will begin to have four CCI cells, but this value is three times less than that of 802.11b systems. The presence of additional channels has another benefit in reducing CCI. As shown in Figure 3.12, the distance between CCI cells is increased, and the likelihood of packets from different cells interfering with one another is therefore reduced.

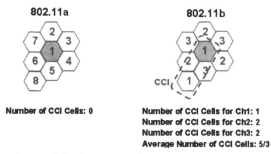

**Figure 3.10.** By virtue of having more channels, 802.11a systems will suffer less CCI than 802.11b systems. Hence, cell throughput will not be degraded in an 802.11a 8-cell system as it will in an 8-cell 802.11b system. Numbers inside each hexagon correspond to different channel frequencies.

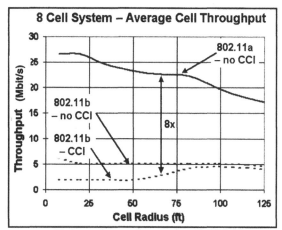

**Figure 3.11.** Average cell throughputs for an 8-cell 802.11a system vs. an 8-cell 802.11b system: 802.11b systems have CCI, and throughput suffers as a result. For a typical cell radius of 65 ft (a diameter of 130 ft), an 802.11a system provides 8 times the average cell throughput of an 802.11b system.

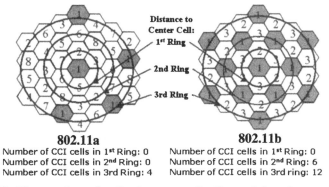

**Figure 3.12.** The number of cells that cause Co-Channel Interference is less for 802.11a systems because of the presence of more channels.

### 3.5.4 Performance and Cost Implications

In Section 3.5.3, we demonstrated how 802.11a can provide more system capacity than 802.11b because of the availability of more channels. This increase allows IT managers to trade off increased performance with lower deployment costs. This is illustrated in Figure 3.13. The total system capacity (average cell throughput multiplied by the number of cells in the system) is plotted against wireless LAN deployment areas for both an eight-cell 802.11b system and different 802.11a systems with a varying number of cells. For a deployment area of 200,000 ft$^2$, a three-cell 802.11a and an eight-cell 802.11b system provide approximately the same system capacity—40.4 and 36.5 Mbit/s, respectively. However, 802.11a can accomplish this with three cells spaced 285 ft apart, whereas an 802.11b system requires eight cells spaced 170 ft apart. This allows 802.11a systems to be provisioned with less AP infrastructure and lower installation costs. Alternatively, IT managers can also choose to deploy an eight-cell 802.11a system to increase system capacity to 158.3 Mbit/s for the same 200,000-ft$^2$ area. In effect, each user now has four times more throughput. Figure 3.13 shows other options that are possible for 802.11a systems.

## 3.6 Conclusions

In this chapter, we have highlighted the benefits of OFDM in dealing with channels with multipath and presented measured 802.11a range performance data in terms of data link rate and throughput. We have used these measurements with an IEEE model to explain the advantages of more frequency channels on system capacity. To summarize, this chapter has produced the following findings:

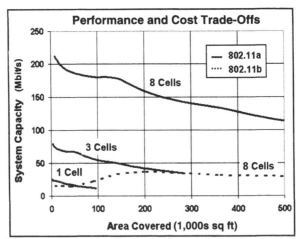

**Figure 3.13.** 802.11a system capacity advantages enable IT managers to deploy the same system capacity using fewer APs than an 8-cell 802.11b system. Alternatively, IT managers can deploy a higher system capacity using the same number of APs as an 8-cell 802.11b system.

- 802.11a has a similar range compared to 802.11b in a typical office environment of up to 225 ft;
- 802.11a has 2–5 times the data link rate of 802.11b in a typical office environment of up to 225 ft;
- 802.11a has 2–4.5 times the throughput of 802.11b in a typical office environment of up to 225 ft;
- 802.11a systems have more available nonoverlapping channels than 802.11b. This allows 802.11a systems to have a higher system capacity compared with 802.11b systems;
- 802.11a has 8 times the system capacity of 802.11b for an 8-cell wireless LAN deployment;
- 802.11a system capacity advantages offer choices for IT network managers. They can either provide the same throughput as 802.11b at lower AP deployment costs or provide increased throughput for similar AP deployment costs as 802.11b.

## References

[1] M. Shoemake and P. Lowry, "802.11b Coexistence Testing Data," IEEE 802.15-010184r0, January 2001.

[2] FCC Part 15 Regulations, February 28, 2001.

[3] K. Ishii, "General Discussion of Throughput Capacity", IEEE 802.11-98, April 23, 1998.

## Acknowledgements

The author expresses special thanks to Jeffrey M. Gilbert and Bill McFarland for their test data, valuable advice, and technical contributions for various sections of this chapter.

Parts of this book have appeared in Communications System Design, Integrated Systems Design, Wireless Systems Design, Electronic Products, and the Atheros Communications White Paper.

## Appendix A: 802.11a Supported Data Rates

| Data Rate (Mbit/s) | IEEE 802.11a Standard Rate | IEEE 802.11a Mandatory Rate | IEEE 802.11a Optional Rate | Modulation and Encoding Rate |
|---|---|---|---|---|
| 6 | Yes | Yes | - | BPSK, R1/2 |
| 9 | Yes | - | - | BPSK, R3/4 |
| 12 | Yes | Yes | - | QPSK, R1/2 |
| 18 | Yes | - | - | QPSK, R3/4 |
| 24 | Yes | Yes | - | 16 QAM, R1/2 |
| 36 | Yes | - | Yes | 16 QAM, R3/4 |
| 48 | Yes | - | Yes | 64 QAM, R2/3 |
| 54 | Yes | - | Yes | 64 QAM, R3/4 |

# Appendix B: Throughput vs. PER Calculation

**Throughput vs. PER**

|  |  |
|---|---|
| sifs (us) = | 16 |
| difs (us) = | 34 |
| slot (us) = | 9 |
| PHY overhead (us) = | 20 |
| data MAC overhead (bytes)= | 30 |
| ACK MAC overhead (bytes) = | 16.75 |
| ACK timeout (greater of actual ack timeout and DIFs time) (us) = | 34 |
| packet error rate = | 0.05 |
| data rate = | 54 |
| packet length = | 1500 |

| try # | probability this try | contention window | ack timeout | time for 1/2 CW | packet time | ACK time (with SIFs) | time this try | accumulated all tries | probability X time |
|---|---|---|---|---|---|---|---|---|---|
| 1 | 9.5E-01 | 15 | 34 | 67.5 | 248 | 40 | 389.5 | 389.5 | 3.70E+02 |
| 2 | 4.8E-02 | 31 | 34 | 139.5 | 248 | 40 | 461.5 | 851 | 4.04E+01 |
| 3 | 2.4E-03 | 63 | 34 | 283.5 | 248 | 40 | 605.5 | 1456.5 | 3.46E+00 |
| 4 | 1.2E-04 | 127 | 34 | 571.5 | 248 | 40 | 893.5 | 2350 | 2.79E-01 |
| 5 | 5.9E-06 | 255 | 34 | 1147.5 | 248 | 40 | 1469.5 | 3819.5 | 2.27E-02 |
| 6 | 3.0E-07 | 511 | 34 | 2299.5 | 248 | 40 | 2621.5 | 6441 | 1.91E-03 |
| 7 | 1.5E-08 | 1023 | 34 | 4603.5 | 248 | 40 | 4925.5 | 11366.5 | 1.69E-04 |
| 8 | 7.4E-10 | 1023 | 34 | 4603.5 | 248 | 40 | 4925.5 | 16292 | 1.21E-05 |
| 9 | 3.7E-11 | 1023 | 34 | 4603.5 | 248 | 40 | 4925.5 | 21217.5 | 7.87E-07 |
| 10 | 1.9E-12 | 1023 | 34 | 4603.5 | 248 | 40 | 4925.5 | 26143 | 4.85E-08 |
| 11 | 9.3E-14 | 1023 | 34 | 4603.5 | 248 | 40 | 4925.5 | 31068.5 | 2.88E-09 |
| 12 | 4.6E-15 | 1023 | 34 | 4603.5 | 248 | 40 | 4925.5 | 35994 | 1.67E-10 |
| 13 | 2.3E-16 | 1023 | 34 | 4603.5 | 248 | 40 | 4925.5 | 40919.5 | 9.49E-12 |
| 14 | 1.2E-17 | 1023 | 34 | 4603.5 | 248 | 40 | 4925.5 | 45845 | 5.32E-13 |
| 15 | 5.8E-19 | 1023 | 34 | 4603.5 | 248 | 40 | 4925.5 | 50770.5 | 2.94E-14 |
| 16 | 2.9E-20 | 1023 | 34 | 4603.5 | 248 | 40 | 4925.5 | 55696 | 1.61E-15 |
| 17 | 1.4E-21 | 1023 | 34 | 4603.5 | 248 | 40 | 4925.5 | 60621.5 | 8.79E-17 |
| 18 | 7.2E-23 | 1023 | 34 | 4603.5 | 248 | 40 | 4925.5 | 65547 | 4.75E-18 |
| 19 | 3.6E-24 | 1023 | 34 | 4603.5 | 248 | 40 | 4925.5 | 70472.5 | 2.55E-19 |
| 20 | 1.8E-25 | 1023 | 34 | 4603.5 | 248 | 40 | 4925.5 | 75398 | 1.37E-20 |
| Total Odds | 1.00 | | | | | | Total Time | | 4.14E+02 |
| | | | | | | | Throughput (Mbps) | | 29.0 |

# Appendix C: Throughput vs. Data Rate and PER Summary

**Throughput vs. Data Rate and Packet Error Rate**

| PER | Data Rate | | | | | | | |
|---|---|---|---|---|---|---|---|---|
| | 6 | 9 | 12 | 18 | 24 | 36 | 48 | 54 |
| 0.00 | 5.4 | 7.8 | 10.1 | 14.2 | 17.7 | 23.9 | 28.7 | 30.8 |
| 0.05 | 5.1 | 7.4 | 9.6 | 13.4 | 16.7 | 22.6 | 27.0 | 29.0 |
| 0.10 | 4.8 | 7.0 | 9.0 | 12.6 | 15.7 | 21.2 | 25.3 | 27.1 |
| 0.15 | 4.6 | 6.6 | 8.5 | 11.8 | 14.7 | 19.7 | 23.6 | 25.2 |
| 0.20 | 4.3 | 6.2 | 7.9 | 11.0 | 13.7 | 18.3 | 21.7 | 23.2 |
| 0.25 | 4.0 | 5.7 | 7.3 | 10.2 | 12.6 | 16.8 | 19.9 | 21.2 |
| 0.30 | 3.7 | 5.3 | 6.8 | 9.4 | 11.5 | 15.2 | 17.9 | 19.0 |
| 0.35 | 3.4 | 4.9 | 6.2 | 8.5 | 10.4 | 13.6 | 15.9 | 16.8 |
| 0.40 | 3.1 | 4.4 | 5.6 | 7.6 | 9.2 | 11.9 | 13.7 | 14.5 |
| 0.45 | 2.9 | 3.9 | 4.9 | 6.6 | 8.0 | 10.1 | 11.6 | 12.2 |
| 0.50 | 2.5 | 3.4 | 4.3 | 5.7 | 6.7 | 8.4 | 9.5 | 9.9 |
| 0.55 | 2.1 | 2.9 | 3.6 | 4.7 | 5.5 | 6.7 | 7.5 | 7.8 |
| 0.60 | 1.8 | 2.4 | 3.0 | 3.8 | 4.3 | 5.2 | 5.7 | 5.9 |
| 0.65 | 1.5 | 2.0 | 2.4 | 2.9 | 3.3 | 3.8 | 4.2 | 4.3 |
| 0.70 | 1.2 | 1.5 | 1.8 | 2.2 | 2.4 | 2.8 | 3.0 | 3.0 |
| 0.75 | 0.9 | 1.2 | 1.4 | 1.6 | 1.7 | 1.9 | 2.0 | 2.1 |
| 0.80 | 0.7 | 0.9 | 1.0 | 1.1 | 1.2 | 1.3 | 1.4 | 1.4 |
| 0.85 | 0.5 | 0.7 | 0.7 | 0.8 | 0.9 | 1.0 | 1.0 | 1.0 |
| 0.90 | 0.4 | 0.5 | 0.6 | 0.7 | 0.7 | 0.7 | 0.8 | 0.8 |
| 0.95 | 0.5 | 0.5 | 0.6 | 0.7 | 0.7 | 0.7 | 0.8 | 0.8 |

# Chapter 4

# Migration Strategies for IEEE 802.11 Wireless LANs

*Proxim, Inc.*

The Institute of Electrical and Electronics Engineers (IEEE) has developed 802.11a, a new specification that represents the next generation of enterprise-class wireless LANs. Among the advantages it has over current technologies are greater scalability, better interference immunity, and significantly higher speed (up to 54 Mbit/s and beyond), which simultaneously allows for higher-bandwidth applications and more users. This chapter provides an overview, in basic terms, of how the 802.11a specification works and its corresponding benefits. It then discusses how existing wireless LAN technologies can be migrated to the 802.11a standard.

## 4.1 The IEEE 802.11a Physical Layer

### 4.1.1 The 5-GHz Frequency Band

802.11a utilizes 300 MHz of bandwidth in the 5-GHz Unlicensed National Information Infrastructure (U-NII) band. Although the lower 200 MHz is physically contiguous, the FCC has divided the total 300 MHz into three distinct 100-MHz domains, each with a different legal maximum power output (Fig. 4.1). The "low" band operates from 5.15 to 5.25 MHz, and has a maximum of 50 mW. The "middle" band is located from 5.25 to 5.35 MHz, with a maximum of 250 mW. The "high" band utilizes 5.725–5.825 MHz, with a maximum of 1 W. Because of the high power output, devices transmitting in the high band will tend to be building-to-building products. The low and medium bands are more suited to in-building wireless products. One requirement specific to the low band is that all devices must use integrated antennas.

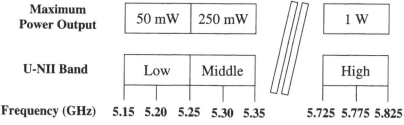

**Figure 4.1.** Specifications of the U-NII 5-GHz frequency band.

Different regions of the world have allocated different amounts of spectrum, so geographic location will determine how much of the 5-GHz band is available. In the United States, the FCC has allocated all three bands for unlicensed transmission. In Europe, however, only the low and middle bands are free. Although 802.11a is not yet certifiable in Europe, efforts are currently under way between IEEE and the European Telecommunications Standards Institute (ETSI) to rectify this. In Japan, only the low band may be used. This will result in more contention for signal but will still allow for very high performance.

The frequency range used currently for most enterprise-class unlicensed transmissions, including 802.11b, is the 2.4-GHz Industrial, Scientific, and Medical (ISM) band. This highly populated band offers only 83 MHz of spectrum for all wireless traffic, including cordless phones, building-to-building transmissions, and microwave ovens. In comparison, the 300 MHz offered in the U-NII band represents a nearly fourfold increase in spectrum, all the more impressive when considering that there is limited wireless traffic in the band today.

### 4.1.2 OFDM Modulation Scheme

802.11a uses Orthogonal Frequency Division Multiplexing (OFDM), a new encoding scheme that offers benefits over spread spectrum in channel availability and data rate. Channel availability is significant because the more independent channels that are available, the more scalable the wireless network becomes. The high data rate is accomplished by combining many lower-speed subcarriers to create one high-speed channel. 802.11a uses OFDM to define a total of eight nonoverlapping 20-MHz channels across the two lower bands; each of these channels is divided into 52 subcarriers, each approximately 300 kHz wide (Fig. 4.2). By comparison, 802.11b uses three nonoverlapping channels.

A large (wide) channel can transport more information per transmission than a small (narrow) one. As described above, 802.11a utilizes channels that are 20 MHz wide, with 52 subcarriers contained within. The subcarriers are transmitted in "parallel," meaning that they are sent and received simultaneously. The receiving device processes these individual signals, each one representing a fraction of the total data that, together, make up the actual signal. With this many subcarriers comprising each channel, a tremendous amount of information can be sent at once.

With so much information per transmission, it obviously becomes important to guard against data loss. Forward Error Correction (FEC) was added to the 802.11a specification for this purpose (FEC does not exist in 802.11b). At its simplest, FEC consists of sending a secondary copy along with the primary information. If part of the primary information is lost, insurance then exists to help the receiving device recover (through sophisticated algorithms) the lost data. This way, even if part of the signal is lost, the information can be recovered so that the data is received as intended, eliminating the need to retransmit. Because of its high speed, 802.11a can accommodate this overhead with negligible impact on performance.

Another threat to the integrity of the transmission is multipath reflection, also called delay spread. When a radio signal leaves the "sending" antenna, it radiates outward, spreading as it travels. If the signal reflects off a flat surface, the original signal and the reflected signal may reach the "receiving" antenna simultaneously. Depending on how the signals overlap, they can either augment each other or cancel each other out (Fig. 4.3). A baseband processor, or equalizer, unravels the divergent signals. However, if the delay is long enough, the delayed signal spreads into the next transmission. OFDM specifies a slower symbol rate to reduce the chance that a signal will encroach on the following signal, minimizing multipath interference.

### 4.1.3 Data Rates and Range

Devices utilizing 802.11a are required to support speeds of 6, 12, and 24 Mbit/s. Optional speeds go up to 54 Mbit/s but will also typically include 48, 36, 18, and 9 Mbit/s. These differences are the result of implementing different modulation techniques and FEC levels. To achieve 54 Mbit/s, a mechanism called 64-level Quadrature Amplitude Modulation (64-QAM) is used to pack the maximum amount of information possible (allowable by the standard) on each subcarrier. Just as with 802.11b, as an 802.11a client device travels farther from its access point, the connection will remain intact but speed decreases (falls back). As Figure 4.4 illustrates, at any range, 802.11a has a significantly higher signaling rate than 802.11b.

## 4.2 The IEEE 802.11a Medium Access Control Layer

802.11a uses the same Medium Access Control (MAC) layer technology as 802.11b, Carrier Sense Multiple Access with Collision Avoidance (CSMA/CA). CSMA/CA is a basic protocol used to avoid signals colliding and canceling each other out. It works by requesting authorization to transmit for a specific amount of time before sending information. The sending device broadcasts a Request to Send (RTS) frame with information on the length of its signal. If the receiving device permits it at that moment, it broadcasts a Clear to Send (CTS) frame. Once the CTS goes out, the sending machine transmits its information. Any other sending devices in the area that "hear" the CTS realize another device will be transmitting and allow that signal to go out uncontested.

## 4.3 Relation to HiperLAN2

HiperLAN2 is a wireless specification developed by ETSI, and it has some similarities to 802.11a at the physical layer. It also uses OFDM technology and operates in the 5-GHz frequency band. The MAC layers are different, however. Whereas 802.11a uses CSMA/CA, HiperLAN2 uses Time Division Multiple Access (TDMA).

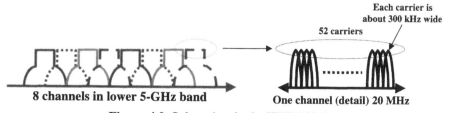

**Figure 4.2.** Subcarriers in the IEEE 802.11a.

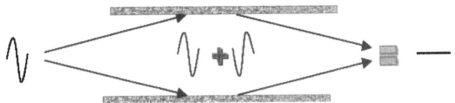

**Figure 4.3.** Multipath interference occurs when reflected signals cancel each other out.

Because the 5-GHz U-NII equivalent bands have been reserved for HiperLAN2 systems in Europe, 802.11a is not yet certifiable in Europe by ETSI. In an effort to rectify this, two additions to the IEEE 802.11a specification have been proposed to allow 802.11a and HiperLAN2 to coexist. Dynamic Channel Selection (DCS) and Transmit Power Control (TPC) allow clients to detect the most available channels and use only the minimum output power necessary if interference is evident. The implementation of these additions will significantly increase the likelihood of European 802.11a certification.

## 4.4 Compatibility with 802.11b

Although 802.11a and 802.11b share the same MAC layer technology, there are significant differences at the physical layer: 802.11b, using the ISM band, transmits in the 2.4-GHz range, whereas 802.11a, using the U-NII band, transmits in the 5-GHz range. Because their signals travel in different frequency bands, one significant benefit is that they will not interfere with each other. A related consequence, therefore, is that the two technologies are not compatible. There are various strategies for migrating from 802.11b to 802.11a or even using both on the same network concurrently.

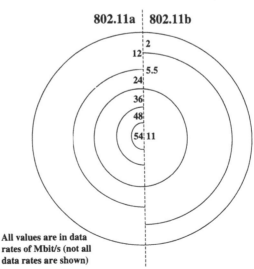

**Figure 4.4.** Example of range/data rate differences between 802.11a and 802.11b.

### 4.4.1 Traditional Access Point Design

Wireless LAN manufacturers that have announced plans for 802.11a products have taken significantly different approaches in addressing the migration issue. The first approach is simply an 802.11a version of a traditional access point, in which each self-contained unit includes the radio, bridging circuitry, processors, and memory required to deliver and manage all wireless-to-wired networking functionality. In this case, 802.11a access points will likely be at a price point equivalent to or greater than traditional 802.11b access points. As discussed earlier, incorporating "traditional-style" 802.11a access points into an existing 802.11b installation will more than double the cost of the complete wireless infrastructure, particularly because a greater number of 802.11a access points will be required to cover the same area at the highest data rate.

### 4.4.2 Slotted Access Point Design

A second approach, which was also attempted in the migration from 2 Mbit/s 802.11 radios to 11 Mbit/s 802.11b radios, is based on access points with a so-called "slotted" architecture. These access points have a slot (or sometimes two slots) into which the radio, in the form of a standard PC card, is inserted. This approach seems to offer advantages in being able to upgrade from one radio technology to the next by simply unplugging the 802.11b PC card and replacing it with a new 802.11a PC card. With the two-slot versions of these access points, the implication is that you can install both radio types side by side within the same unit.

However, the flaw in this approach lies in the significant differences in data rate versus range associated with each radio technology, as discussed previously. The location of each existing 802.11b access point in any given installation will have been determined from a site survey based on the range of the 802.11b radio. Because access points are positioned to provide a maximum of coverage with a minimum of units, there is typically little overlap within the coverage area. If the 802.11b radio is simply removed and replaced with an 802.11a radio, the location of the preexisting access points will not be optimized for the maximum data rate offered by 802.11a. Therefore, there will be areas between the "new" 802.11a access points in which only a reduced (backoff) data rate will be available. To achieve optimal performance of the 802.11a installation, a new site survey based on 802.11a radios will be necessary, which will dictate that the existing access points be relocated and additional 802.11a access points be installed. The costs associated with such relocation and addition of equipment must be considered.

### 4.4.3 A Multistandard Wireless LAN Architecture

A more intelligent approach is that taken by Proxim with its new multistandard Harmony architecture. Harmony introduces a new concept in wireless LAN infrastructure whereby much of the cost has been removed from the access points, with all networking and management functions being incorporated into a new device called an Access Point Controller. The Harmony Access Point Controller and associated Harmony Access Points communicate with each other over the existing wired infrastructure. Currently, Harmony 802.11b Access Points are available, with equally low-cost Harmony 802.11a Access Points planned for introduction soon (Fig. 4.5).

The new Harmony architecture addresses the issues of cost, coverage, and management in a unique way. The quantity and location of each low-cost 802.11a access point will be based on an appropriate 802.11a site survey. The total cost of the entire 802.11b *and* 802.11a wireless network infrastructure will be significantly lower than with traditional or slotted access points. Additionally, because all wireless LAN management functions are centralized in the Harmony Access Point Controller, administrators are presented with a unified management interface for all access points, regardless of radio technology. With the Harmony system, configuration information for *all* access points (e.g., channelization, packet filtering, security parameters, etc.) is maintained within the Access Point Controller.

Installation of access points, whether 802.11b or 802.11a, is a simple matter of plugging the access point into the existing wired infrastructure; the Access Point Controller will automatically "discover" the new access point and download all configuration parameters. Upgrades and maintenance functions are equally simple, because changes are entered *once* in the Access Point Controller and automatically distributed to each relevant access point.

## 4.5 Conclusions

802.11a represents the next generation of enterprise-class wireless LAN technology, with many advantages over current options. At speeds of 54 Mbit/s and greater, it is faster than any other unlicensed solution. 802.11a and 802.11b both have a similar range, but 802.11a provides higher speed throughout the entire coverage area. The 5-GHz band in which it operates is not highly populated, so there is less congestion to cause interference or signal contention. In addition, the eight nonoverlapping channels allow for a highly scalable and flexible installation. Given these advantages, it is important to examine how existing wireless LAN technologies can be migrated to the 802.11a standard. A cost-effective method of achieving this is to employ a centralized wireless LAN management controller by which administrators are presented with a unified management interface for all access points, regardless of radio technology.

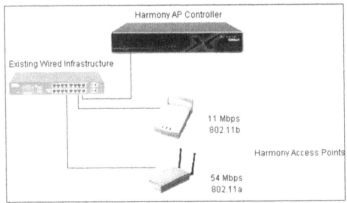

**Figure 4.5.** Proxim's patented Harmony architecture.

# Chapter 5

# 5-GHz Radio Spectrum Regulations

Teik-Kheong Tan
*3Com Corporation, Inc.*

The 5-GHz spectrum for licensed exempt devices is *exclusively* available in the US, Canada, Japan, and Europe (HiperLAN devices only). Australia is now starting to consider making an assignment. The biggest challenge for the 5-GHz wireless LAN industry is that the current permitted power in the 5-GHz spectrum is being challenged by primary users. Primary users, in particular users from the satellite industry, have a very experienced lobbying machine compared with the highly disorganized IT industry. The ITU-R has included an agenda item in World Radiocommunication Conference 2003 (WRC 2003) for harmonization of the spectrum. They also have two agenda items for studies on spectrum sharing between mobile devices and primary users. This chapter addresses the current issues and regulatory aspects of the 5-GHz spectrum allocation in the US, Europe, and the rest of the world.

## 5.1 Introduction

The 5-GHz band has become the new frontier for high-bandwidth wireless LAN products. Being spectrally clean and wide, the 5-GHz band attracts much attention as the key enabler of public acceptance for broadband wireless LAN products. Recent and ongoing development of wireless LAN standards (e.g., IEEE 802.11a [1], ETSI HiperLAN) aimed at the interoperability between wireless LAN devices produced by different manufacturers is expected to increase market demand for wireless LAN devices. The prospect of license-free operation of wireless LANs makes them even more attractive to end users because it reduces the cost of operation and administrative overhead in the establishment of low-cost wireless networking. The issue faced today by the entire 5-GHz wireless LAN industry is the sharing of 5-GHz spectrum with primary users. The primary users include:

- Government radiological systems;
- Mobile satellite feeder links;
- Amateur operations;
- Industrial, scientific, medical (ISM) operations;
- Other unlicensed FCC Part 15 operations;

- Proposed Intelligent Transportation Systems.

The overall 5-GHz bandwidth allocation is divided into the following bands:

- The 5.0- to 5.250-GHz band is allocated on a primary basis to the aeronautical radionavigation and aeronautical mobile satellite, fixed satellite, and intersatellite services for both government and nongovernment operations;
- The 5.250- to 5.350-GHz band is allocated to the nongovernment radiolocation service on a secondary basis;
- The 5.650- to 5.925-GHz band is allocated on a secondary basis to the amateur service;
- The 5.725- to 5.875-GHz band is designated for ISM applications and unlicensed Part 15 devices, and radiocommunication services operating within this band must accept harmful interference that may be caused by ISM applications;
- The 5.850- to 5.925-GHz band is allocated on a primary basis to the fixed satellite (Earth to space) service for nongovernment operations and to the radiolocation service for government operations.

Currently, the IEEE 802.11a and HiperLAN2 are two leading 5-GHz standards for wireless LANs. The IEEE 802.11 committee finalized the 802.11a standard using a new OFDM-based PHY layer and the original 802.11 MAC layer. Current work in several IEEE 802.11 task groups focuses on MAC enhancements, to allow quality of service, advanced privacy, and various radio control functionalities. The ETSI BRAN committee is finalizing the HiperLAN2 standard, sharing the same OFDM PHY layer philosophy with the 802.11 (actual implementation is slightly different) and incorporating an advanced, feature-rich MAC layer (DLC in the BRAN terminology). There are a number of organizations worldwide developing standards for wireless LANs. The big three are ETSI BRAN, the IEEE 802.11, and MMAC in Japan. Their joint participation resulted in a 1998 harmonization effort on the wireless PHY layer at 5 GHz. This first step resulted in partial harmonization of the standards and promoted coexistence.

As radio frequencies in the 5.150- to 5.875-GHz bands are becoming available on a global basis for wireless LANs, there is increasing urgency for the wireless network industry to work with the regulatory bodies to ensure that spectrum issues with the primary services are resolved.

## 5.2 Spectrum Regulations Today

The available spectrum at 5 GHz is currently defined within the ITU-R Radio Regulations for a number of different services. However, WRC 2000 agreed that the 5-GHz band be considered for allocation to Mobile Service at WRC 2003, providing an opportunity to harmonize, on a global basis, the spectrum available for use by wireless LAN devices. It is important to note that there are currently differences among CEPT, Australia, Japan, the US, and Canada with respect to EIRP levels, frequency bands, and operational restrictions.

### 5.2.1 Regulatory Arrangements in Europe

Within Europe, the 5.150- to 5.350-GHz and 5.470- to 5.725-GHz bands have been designated for use by HiperLANs. For the 5.150- to 5.350-GHz band, the maximum mean EIRP is 200 mW and the band is restricted to indoor usage because it is to be shared with Fixed Satellite Service (FSS). For the 5.470- to 5.725-GHz band, the maximum mean EIRP is 1 W and both indoor and outdoor usage is possible. Furthermore, Dynamic Frequency Selection is mandatory to provide a uniform loading of HiperLANs across a minimum bandwidth of 330 MHz (or 255 MHz in the case of equipment used in the 5.470- to 5.725-GHz band). In addition, transmit power control ensures a mitigation factor of at least 3 dB for both uplink and downlink.

### 5.2.2 Regulatory Arrangements in the US

In January 1997, the FCC made a decision concerning Unlicensed National Information Infrastructure (U-NII) allocation in the 5-GHz band. In the FCC's rules for U-NII devices, a U-NII device is defined as [2]:

*Intentional radiators operating in the 5.15–5.35 GHz and 5.725–5.825 GHz frequency bands that use wideband digital modulation techniques and provide a wide array of high data rate mobile and fixed communications for individuals, businesses, and institutions*

Furthermore, the FCC believes that U-NII devices [3]

*will provide short-range, high-speed wireless digital communications and may offer new opportunities for providing advanced telecommunications services to educational institutions, health care providers, libraries, businesses, and other users*

The total U-NII bandwidth allocation provides 300 MHz of unlicensed spectrum at the 5.15- to 5.35-GHz and 5.725- to 5.825-GHz bands for short-range, high-speed wireless communications. For the 5.150- to 5.250-GHz band, the power limit is 50 mW for indoor usage only. This first band is to be used for short-range wireless communications such as communication between computers and printers within a fixed network. This will typically be in a room or in adjoining rooms. Because operation is restricted to the indoors, it will be difficult for U-NII devices and Mobile Satellite Service (MSS) feeder links to interfere with each other. In addition, walls and ceilings provide protection against interference due to attenuation effects.

For the 5.250- to 5.350-GHz band, the power limit is 250 mW for both indoor and outdoor usage. Outdoor operations are permitted because it will not be shared with the MSS. The 250-mW peak transmitter output power will be sufficient for communication within and between buildings, as in a campus LAN.

Finally, for the 5.725- to 5.825-GHz band, the power limit is 1 W. In this band, devices are already allowed to operate at 1-W peak transmitter output power within current regulations. These devices typically use high-gain antennas to construct community networks that span 1–2 km.

### 5.2.3 Regulatory Arrangements in Japan

In Japan, only the 5.150- to 5.250-GHz band is currently allocated for wireless LAN operation. The following RF center frequencies are specified: 5.170, 5.190, 5.210, 5.230 GHz. The transmission rate is equal or greater than 20 Mbit/s, and the transmission power is less than 10 mW/MHz. To facilitate frequency sharing, Japan requires that the spectrum is not monopolized by a single 5-GHz system. Therefore, a carrier sense method is required to allow users to detect the presence of any ongoing transmissions. The effective period of the carrier sensing is 4 ms with a sense level (the electric field strength) of 100 mV/m.

### 5.2.4 Regional Regulations

Table 5.1 (based on [4]) summarizes the spectrum allocation in various regions throughout the world.

## 5.3 Issues with Spectrum Regulations Today

Considerable work, including spectrum sharing, has been initiated by ITU-R's study forums on wireless LANs. In this case, HiperLAN characteristics were used extensively to study spectrum-sharing arrangements between wireless LANs and primary services. In a corresponding study in the US, the FCC's Final Report and Order (R&O) on ET Docket 96-102 recognizes that several primary services are currently using the spectrum proposed for U-NII devices. The ITU-R studies were unfortunately conducted without the US specifications. In this section, two broad issues related to the usage of the 5-GHz spectrum are discussed.

### 5.3.1 Sharing with Primary Users

The basic issue arises from the WRC primary allocation to satellite and radiolocation services. There have been continuous debates in both the US and Europe regarding the interference from wireless LAN operations on the primary services. For example, in the 5.150- to 5.250-GHz band, parties with MSS interests have argued that sharing is not feasible between MSS feeder links and new U-NII devices. In particular, they assert that it will take only 1,070 simultaneous users of U-NII devices operating in that band to cause unacceptable interference to the MSS feeder links. On the other hand, U-NII proponents claim that U-NII devices operating in the 5.150- to 5.250-GHz band will be able to share with government radiolocation and MSS feeder uplink operations. This is because of the low power at which U-NII devices operate and the high attenuation characteristics of radio signals in the 5-GHz frequency band.

With regard to sharing with MSS feeder uplink operations, U-NII proponents conclude that MSS feeder operations will also be able to share with U-NII devices. They base this conclusion on the ITU study, which predicted that HiperLAN systems will be able to coexist with MSS feeder links in the 5.150- to 5.250-GHz band in Europe. Furthermore, they note that HiperLAN devices will be operating at

1 W, a power level substantially higher than the power limit proposed for U-NII devices in this band, and that global MSS systems must be robust enough to share with HiperLAN devices.

### Table 5.1. Worldwide 5-GHz spectrum allocation.

| Region | (A1) 5.150–5.250 GHz | (A2) 5.250–5.350 GHz | (B) 5.470–5.725 GHz | (C) 5.725–5.875 GHz (ISM) |
|---|---|---|---|---|
| USA | U-NII band FCC Part 15 subpart E | U-NII band FCC Part 15 subpart E | Not allowed | U-NII band FCC Part 15 subpart E |
| Power limit | 50 mW Indoor only | 250 mW Indoor/Outdoor | | 1 W Indoor/Outdoor |
| Power limit EIRP | 200 mW | 1 W | | 4 W (FWA) 200 W (point to point with highly directional antennas) |
| Licensing | Unlicensed | Unlicensed | | Unlicensed |
| Coexistence/E tiquette | None | None | | None |
| Canada | RSS-210 Issue 3 Par. 6.2.2 (q1) | RSS-210 Issue 3 Par. 6.2.2 (q1) | Not allowed | RSS-210 Issue 3 Par. 6.2.2 (q1) |
| Power limit | None stated Indoor only | 250 mW | | 1 W |
| Power limit EIRP | 200 mW | 1 W | | 4 W (FWA) 200 W (point to point with highly directional antennas) |
| Licensing | Unlicensed | Unlicensed | | Unlicensed |
| Coexistence/E tiquette | None | None | | None |
| Europe | ERC Decision (99)23 | ERC Decision (99)23 | ERC Decision (99)23 | ISM |
| Power limit EIRP | 200-mW HiperLAN Indoor only | 200-mW HiperLAN Indoor only | 1-W HiperLAN Indoor/outdoor | 25 mW |
| Licensing | License exempt | License exempt | License exempt | Unlicensed |
| Coexistence/E tiquette | 20-MHz channels assigned DFS & TPC mandatory | 20-MHz channels assigned DFS & TPC mandatory | 20-MHz channels assigned DFS & TPC mandatory | - |
| Japan | | FWA not allowed | Not allowed | Unknown |
| Power limit EIRP | 200 mW | | | |
| Licensing | Unlicensed | Licensed (FWA) | | |
| Coexistence/E tiquette | None | | | |
| Australia | SP 1/00—May 2000 Radiocommunications Class License (Low-Interference-Potential Devices) 2000 | SP 1/00—May 2000 Radiocommunications Class License (Low-Interference-Potential Devices) 2000 | | SP 1/00—May 2000 Radiocommunications Class License (Low-Interference-Potential Devices) 2000 |
| Power limit EIRP | 200 mW Indoor only | 200 mW Indoor only | | 1 W |
| Licensing | Class license | Class license | | Class license |
| Coexistence/E tiquette | None | None | | None |

In another report, ITU-R JRG 8A-9B studied the sharing of wireless LAN devices and Non-Geostationary Orbit (NGSO) MSS feeder links in the 5.150- to 5.250-GHz band. This has resulted in the adoption of a Draft New Recommendation (DNR) on EIRP density limits for wireless LANs in this band (ITU-R 8/95) [5]. The mean EIRP density limit recommended is 10 mW/MHz, with an overall limit of 200 mW per device. Furthermore, the DNR stipulates that devices are restricted to indoor use only. The DNR also recommends that Transmit Power Control (TPC) and spectral spreading Dynamic Frequency Selection (DFS) techniques be considered as additional mitigation techniques. In addition, recent work reported in ITU-R 7C/158 Attachment 9 [6], ITU-R 7C/174 [7], and ITU-R 7C/192 [8] in regard to sharing among wireless LANs, EESS, and SRS recommended that wireless LAN operation in the 5.250- to 5.350-GHz frequency range be restricted to indoor use with an EIRP of 200 mW per device. The ITU-R documents also suggested that TPC and DFS be used to facilitate sharing.

In the 5.725- to 5.825-GHz band, incumbent operators either oppose allowing U-NII operation because of interference concerns or urge that sharing studies be completed before that band is made available to U-NII devices. Furthermore, these operators oppose high-power U-NII operation in this band because it is likely to cause more interference. On sharing between U-NII devices and amateur operations, U-NII proponents argue that the proposed U-NII maximum power limit of 100 mW EIRP is sufficient to avoid significant interference to the amateur service. Opponents, however, state that the ubiquitous nature, mobility, and aggregate interference potential of these devices necessitate that sharing studies be performed. Additionally, FCC Rules Part 15 (Section 15.247) spread spectrum interests oppose U-NII operations in this band and argue that without a means to control usage, operation in this band will rapidly degrade and become unusable.

In summary, there has been no conclusive argument against interference generated by 5-GHz wireless LANs. Every debate has been rife with inconsistencies, and there seems to be no end to the claims made by both sides.

### 5.3.2 Additional Issues

The other issues in sharing the 5-GHz spectrum are:

- Estimation of spectrum need;
- Global harmonization;
- Coexistence between different wireless LAN standards;
- Fixed wireless access;
- Emerging residential market.

Not enough studies have been done to estimate the spectrum needs for wireless LANs. Spectrum estimation should take into account the amount of sharing, the actual bandwidth consumption by services, and the impact of coexistence of different standards. There have been strong requests for the adoption of a methodology similar to 3G services to be used in the estimation of spectrum.

As to global harmonization, the 5-GHz Wireless Industrial Advisory Group (5GWIAG) started by Microsoft, Intel, and Compaq initiated a new harmonization effort in July 2000. The group requested ETSI BRAN, IEEE, and MMAC to collaborate and achieve a single unified standard. Recent approval by IEEE of the formation of the IEEE 802.16.4 committee named WirelessHUMAN has also laid claim to the nascent 5-GHz band. The focus of this group so far has been to examine PMP and mesh-type FWA topologies with a PHY layer derived from the 802.11a and HiperLAN2 standards, with mandatory elements such as TPC and DFS enabling coexistence with the current two standards.

Finally, the emerging residential market has spawned many new requirements for 5-GHz operation in the home. Adequate spectrum management and coexistence measures for high service quality are among the top priorities for residential services. The primary users are concerned, given the potentially high user density in a residential neighborhood coupled with intensive 5-GHz emission per household. The aggregate emission may eventually exceed the EESS interference threshold.

## 5.4 Conclusions

It is likely that a number of interested parties will propose that a future WRC consider making spectrum available on a worldwide basis for 5-GHz wireless LANs. For instance, the IEEE, in its document IEEE 802.11-99/264 [9], advocates the need for worldwide 5-GHz allocation by a WRC for wireless LANs. Furthermore, at the next WRC in 2003, an agenda item seeking harmonized spectrum for mobiles within the 5-GHz bands is requested. The Mobile Service allocation was requested in ECP by CEPT. Agenda Item 1.5 (from JTG 4-7-8-9: WRC-2003) reads:

*to consider, in accordance with Resolution [GT PLEN-2/1] (WRC-2000), regulatory provisions and spectrum requirements for new and additional allocations to the mobile, fixed, earth exploration-satellite and space research services, and to review the status of the radiolocation service in the frequency range 5.150–5.725 GHz, with a view to upgrading it, taking into account the results of ITU-R studies*

It is worth noting that the scope of Agenda Item 1.5 does not include the U-NII ISM band. The ITU-R might also consider discussions on sharing the spectrum in the 5.350- to 5.725-GHz band between wireless LAN and primary users.

For the 5-GHz wireless LAN industry, this is the time to unify and collaborate to address the needs of the industry. Given the fragmented lobbying powers in the general IT industry, it makes sense for companies with investments in either IEEE 802.11a or HiperLAN2 products to be more proactive in regulatory and standards meetings such as the ITU-R, IEEE, ETSI BRAN, and MMAC. There is an urgent need for more sector members from the wireless LAN industry in the ITU-R. Increased participation from many countries in national preparatory meetings for ITU-R meetings will be of great help. Finally, the industry as a whole can benefit from this important exercise by having more RF spectrum skill set available.

## References

[1]    IEEE   802.11a-1999,   "Supplement   to   Information   Technology—Telecommunications and Information Exchange Between Systems—Local and Metropolitan Area Networks—Specific Requirements—Part 11: Wireless LAN Medium Access Control and Physical Layer Specifications: High Speed Physical Layer in the 5 GHz Band," November 1999.

[2]    Code of Federal Regulations, Title 47, Volume 1, Part 15, Sub-part E, Sections 15.401 to 15.407 Unlicensed National Information Infrastructure, 31 July 1998.

[3]    FCC 96-193 Notice of Proposed Rule Making, 6 May 1996.

[4]  Draft Recommendations on the Licensing of the 5 GHz (5.150–5.350, 5.470–5.725, 5.725–5.875 GHz) Frequency Bands, 5GAG.

[5]  ITU-R Document 8/95-E, "Draft New Recommendation ITU-R M. [DOC. 8/95]—EIRP Density Limit and Operational Restrictions for RLANs or Other Wireless Access Transmitters in Order to Ensure the Protection of Feeder Links of Non-Geostationary Systems in the MSS in the Frequency Band 5.150–5.250 GHz," 17 August 1999.

[6]  ITU-R Document 7C/158 Attachment 9, "Preliminary Draft New Recommendation—Sharing in the Bands 5.250–5.570 GHz Between the Earth Exploration-Satellite (Active) and Space Research (Active) Services and Other Services Allocated in This Band," 22 March 1999.

[7]  ITU-R Document 7C/174, "Preliminary Draft New Recommendation—Sharing in the Bands 5.250–5.570 GHz Between the Earth Exploration-Satellite (Active) and Space Research (Active) Services and Other Services Allocated in This Band," 14 January 2000.

[8]  ITU-R Document 7C/192, "Sharing Study Between HiperLAN and Earth Exploration Satellite Systems Relating to Preliminary Draft New Recommendation on Sharing in the Band 5.250–5.570 GHz Between the EESS (Active) and Space Research Service (Active) Service and Other Services in the Band," 18 January 2000.

[9]  IEEE 802.11-99/264, "IEEE P802.11 Wireless LANs, Draft Letter for 5 GHz Coexistence," November 1999.

# Chapter 6

# Quality of Service and Multimedia Support in IEEE 802.11 Standards

Gregory Parks
*Cirrus Logic, Inc.*

As the convergence of data and entertainment has increased in pace, there has been a large amount of corresponding activity within the IEEE 802 standards committees to create enhancements to existing standards that will be capable of supporting this convergence. Much of the focus has been in the IEEE 802.11 Wireless LAN committee, and much of the emphasis there has been on explicit support for Quality of Service (QoS) enhancements and multimedia applications, although other committee activities have proposed enhancements that may also be used to support QoS and multimedia applications. This chapter describes the current state of the proposed changes to the relevant IEEE802.11 standards and provides context for the reasons behind those proposals.

*__Disclaimer__ – It is important to emphasize from the beginning that the author, although presently a voting member of the IEEE 802.11 committee, does not speak for that committee and should not be viewed as speaking for that committee. The views presented herein are the views of the author alone and may or may not coincide either with the views of the official representatives of the committee or with the other individual members of the committee.*

## 6.1 Introduction

The IEEE standards committees consist of individual members who may or may not be sponsored by an organization. The committee meets in plenary session every 4 months and typically meets in interim session in the intervening intervals. The activities consist largely of presentations, deliberations, and draft composition on mostly technical and sometimes procedural issues. The sometimes multiyear procedure followed in achieving the goal of an accepted and successful standard begins with the creation of a working draft that can be passed by 75% of the voting membership in an all-inclusive letter ballot. Once passed, this letter ballot is then recirculated with new drafts resolving comments from the voters who did not vote to approve the draft. Ultimately, when a level of 90% approval or more is reached, the draft is submitted to sponsors for their comment and approval, eventually leading to a new release of the standard. This lengthy process operates under IEEE Rules and Robert's Rules of Order, and every effort is made to be sure that each and every voice and opinion has been heard and considered. Current IEEE 802 standards fall into three categories defined by the IEEE. The three network types differ largely by range and by usage. Although the other network layers are described here for completeness, only wireless LANs will be discussed in the remainder of this paper.

### 6.1.1 Wireless LAN—IEEE802.11

The original wireless LAN standard specifies the Medium Access Control (MAC) layer and a number of Physical (PHY) layers. These layers correspond to layer 2 and layer 1 of the Open System Interconnection (OSI) hierarchy. The MAC employs both an Ethernet-like Contention Period (CP)-based medium access scheme and an optional Point Coordination Function (PCF) relying on operation in a Contention-Free Period (CFP) using a polling mechanism alternating in time with the CP access mechanism. The PHY layer was designed to work in the 2.4-GHz band using either Frequency Hopping Spread Spectrum (FHSS) or Direct Sequence Spread Spectrum (DSSS) technology as well as in the optical infrared (IR) band. This standard was approved in 1997 and updated in 1999 and has since become both an IEEE and an ISO standard. It may be argued that this wireless LAN standard is the most widely deployed of all wireless LAN standards, particularly in its IEEE802.11b revision.

- IEEE802.11a—This is a PHY layer specification that extends the bandwidth of the 802.11 network to up to 54 Mbit/s in the 5-GHz band by means of the use of Orthogonal Frequency Division Multiplexing (OFDM). This standard is not meant to be interoperable with any other 802.11 PHY layer standard.

- IEEE802.11b—This is a PHY layer specification that extends the bandwidth of the 802.11 network to up to 11 Mbit/s in the 2.4-GHz band by means of the use of DSSS. This standard is designed to be backward compatible with the 1 Mbit/s and 2 Mbit/s 802.11 base rates defined in the original IEEE802.11-1997. Although the standard is lower in rate than IEEE802.11a, the approval process started later than that for IEEE802.11a, hence the unusual lettering sequence.

### 6.1.2 Wireless Personal Area Network—IEEE802.15

This standard is still under development and is intended to provide IEEE standardization for the specification developed by the "Bluetooth" Special Interest Group (SIG). This committee also includes work on high-rate and low-rate variants of the wireless personal area network.

### 6.1.3 Wireless Metropolitan Area Networks—IEEE802.16

This standard is still under development and is intended to provide primarily wireless point distribution technology on a number of different PHY layers to address the cable distribution system "last mile" problem where transmissions must be carried from some intermediate central point to subscribers.

## 6.2 IEEE 802.11 Standards Task Groups

In keeping with the amount of recent standards activity, there are a number of officially constituted task groups. For completeness, all are listed, although not all are discussed in the context of QoS and multimedia. For accuracy and brevity, the official charter for each task group has been quoted in abridged form.

### 6.2.1 TGd—Regulatory Domains

"The current 802.11 standard defines operation in only a few regulatory domains (countries). This supplement will add the requirements and definitions necessary to allow 802.11 wireless LAN equipment to operate in markets not served by the current standard."

### 6.2.2 TGe—Extensions for Multimedia and QoS

"To enhance the current 802.11 MAC to expand support for applications with Quality of Service [QoS] requirements [as well as] in the capabilities and efficiency of the protocol. These enhancements, in combination with recent improvements in PHY capabilities from 802.11a and 802.11b, will increase overall system performance, and expand the application space for 802.11."

### 6.2.3 TGf—Inter-Access Point Protocol

"To develop recommended practices for an Inter-Access Point Protocol (IAPP) which provides the necessary capabilities to achieve multi-vendor Access Point interoperability across a Distribution System supporting IEEE P802.11 Wireless LAN Links. [...] This Recommended Practice Document shall support the IEEE P802.11 standard revision(s)."

### 6.2.4 TGg—High-Rate 2.4 GHz

"To develop a new PHY [layer] extension to enhance the performance and the possible applications of the 802.11b compatible networks by increasing the data rate achievable by such devices. This technology will be beneficial for improved access to fixed network LAN and inter-network infrastructure (including access to other wireless LANs) via a network of access points, as well as creation of higher performance ad-hoc networks."

### 6.2.5 TGh—Spectrum and Transmit Power Management

"To enhance the current 80.11 MAC and 802.11a PHY [layer] with network management and control extensions for spectrum and transmit power management in 5 GHz license exempt bands, enabling regulatory acceptance of 802.11 5 GHz products. Provide improvements in channel energy measurement and reporting, channel coverage in many regulatory domains, and provide Dynamic Channel Selection and Transmit Power Control mechanisms."

### 6.2.6 TGi—Security

"To enhance the current 802.11 MAC to provide improvements in security." [To avoid any confusion it is important to point out that until recently this task group operated as part of the TGe task group.]

There are other ad hoc groups as well as these formal committees, but these ad hoc groups are formulated as an aid to solving particular issues using experts drawn from the larger committee.

## 6.3 Key Concepts

Before we look at the specifics of the IEEE wireless LAN Working Group proposals, it is important to have a firm understanding of certain key QoS and multimedia concepts as the background and context in which these proposals are being developed.

*Native (MAC) Services*—There are two ways to approach the problem of supporting multimedia and QoS at the MAC layer: either (1) rely on the existence of sufficient bandwidth so that no further support is needed or (2) design support into the MAC for the various types of QoS and the expected traffic types. By way of example, modern switched Ethernet may properly rely on the former, whereas whether wireless LANs must rely on the latter is a topic of continuing debate.

*Differentiated Services (diffserv)*—This allows the network to allocate different levels of service to different traffic on the network. Broadly speaking, any traffic management or bandwidth control mechanism that treats different streams differently—ranging from simple Weighted Fair Queueing to RSVP and per-session traffic scheduling—counts. However, in common usage the term is coming to mean any relatively simple, lightweight mechanism that does not depend entirely on per-flow resource reservation and to indicate that even without explicit parameterization a difference in throughput between different priorities can be discerned, thereby permitting some differentiation between traffic types (such as data vs. voice).

*Integrated Services (intserv)*—This allows the network to support audio, video, voice, and other multimedia traffic using Internet Engineering Task Force (IETF) Service Level Agreements (SLAs) that provide for explicit reservation and parameterization of traffic on a per-stream basis; an intermediate form is available that may support parameterization for traffic on a per-packet rather than per-stream basis with typically all packets of a certain traffic type being identified with the same tag (often, the priority tag) regardless of stream membership. This intermediate form allows for greater control over the differentiation of a range of traffic types than that provided by simple prioritization while avoiding the complexity of stream parameterization. It provides simple separation between traffic types but not separation between streams of the same type, because there is no notion of a stream.

*Application Services*—diffserv and intserv are fine in the roles defined for them, but they do not resolve the entire QoS problem. Without support from the layers above the transport layer, QoS is unlikely to be specified or maintained. This is because the QoS intserv and diffserv protocols rely on the upper layers either to tag the traffic appropriately or to create a path for the stream and specify the path parameters.

*Implicit Services*—This allows the network to support QoS and multimedia applications without explicit parameterization or prioritization (also known as hot QoS). Providing this support is presently important because so few Application Services are actually provided.

*Constrained Bandwidth*—Working in the wireless world has usually meant being constrained by the available bandwidth. Although wired transmissions in LANs have moved to well beyond 100 Mbit/s, with 1 Gbit/s rates available in general practice, wireless LANs remain mired at 11 Mbit/s in general practice with the upper IEEE-standardized limit being 54 Mbit/s over short distances. This bandwidth constraint translates into the need to conserve bandwidth under conditions of high utilization. In addition, multimedia traffic is different from ordinary traffic by virtue of the fact that the bandwidth of the originator is typically constrained by design, whereas nonmultimedia traffic is not usually explicitly constrained (although it is implicitly constrained by equipment speeds).

*Reserved Bandwidth*—Sometimes referred to as guaranteed bandwidth, this refers to the concept that bandwidth asked for by a stream will be available whenever the stream needs it, usually specified in peak, minimum, and average bandwidth.

*Bounded Latencies*—Sometimes referred to as guaranteed latencies, this refers to the concept that data will be delivered with an upper limit on the amount of latency present in that delivery, usually specified in maximum latency, with some amount of tolerance in the spread of latencies (jitter) that will be encountered.

*Admittance Control*—Because bandwidth is constrained, it may be necessary to reject the admittance of additional traffic to preserve outstanding connection agreements.

*Traffic Types*—A number of different multimedia traffic types, each with differing characteristics, can be identified in current applications:

- Best Effort—Although not strictly a multimedia type, it must be mentioned because most nonmultimedia traffic as well as much current multimedia traffic is of this type; no guarantees of any kind are made with respect to bandwidth, latency, or even eventual delivery, only that the network will do its best to successfully deliver this type of traffic to the destination.
- Audio—Either PCM audio or audio compressed with a coder such as MP3, this traffic is of medium rate and exhibits a very low tolerance to reception errors; any missing data is easily detected by ear, and data arriving late must be discarded.
- Video—Typically compressed with a coder such as MPEG1 or MPEG2, video is of high rate and may be very sensitive to the latencies determined by the way the decoder works, often sensitive to both overall latency as well as latency jitter: both must be kept to a minimum to minimize receive buffering; missing data results in a visibly static picture until reception of the next sequential key frame occurs.
- Interactive Gaming—Typically of low rate but very sensitive to the real-time latency experienced between the event and the arrival of that event at the destination and often coupled to a return path such as a visual event; there is no

defined latency bound, but the longer the latency the slower and less competitive the player.

- Voice, VoIP, and Video Conferencing—Of low to medium rate but very sensitive to end-to-end bidirectional latency; the rate and the tolerance to reception errors are dependent on the type of coder(s) being used, but the type of coder cannot make up for excess latencies.
- IEEE1394 "FireWire"—Very high rate and very sensitive to uninterrupted isochronous operation.

*An Example*—Consider the situation in which an unconstrained best effort print stream, a voice stream, and a 6 Mbit/s MPEG2 video stream are all moving simultaneously over a 10 Mbit/s wireless LAN segment (or wired LAN for that matter). In isolation, each of the streams will be handled correctly. But when the unconstrained print stream is added to the wireless LAN, statistically the bandwidth available for the video stream will be less than ½ of the 10 Mbit/s available at the medium and the video stream will be subsequently degraded by lack of adequate bandwidth. Meanwhile, the voice stream, although of low bandwidth, must contend with both the video and print streams for channel access, likely increasing latency. Ideally, the video stream would have reserved bandwidth sufficient for its needs and would be protected by admittance control from similar streams, whereas the print and voice streams, with priority given to the voice stream, would only contend with each other.

## 6.4 TGe—Multimedia and QoS

Now we can get into the specifics of the proposed changes to the standard, moving from the most general to the most specific enhancement. This is not meant to be a comprehensive list of proposed changes and only covers the most pertinent changes.

*General Enhancements*—These enhancements affect the overall operation of the MAC:

- STA-STA—The current IEEE802.11 standard permits transmitting from one Station (STA) to another only in ad hoc mode. Most networks are operated in infrastructure mode, where an Access Point (AP) serves as the originator or destination for all traffic within a Basic Service Set (BSS, or subnet) acting as an intermediary between STAs in the same BSS. This is very inefficient in terms of channel usage and in terms of absolute bandwidth because each STA-to-STA stream must pass through the AP. The TGe proposal includes the ability for STA-to-STA transmission to bypass the AP and move instead directly between the relevant STAs within the same BSS.

- Negotiable Acknowledgments—The existing IEEE 802.11 standard relies on a positive acknowledgment mechanism for all frame exchanges. The TGe proposals introduce the notion of negotiable acknowledgments whereby a packet in a stream need not be acknowledged or such acknowledgments can be aggregated depending on the parameterization of a stream. Such negotiable acknowledgments lead to a more efficient utilization of the available channel and enable for such applications a reliable multicast.

- Forward Error Correction (FEC)—FEC provides for the ability to encode a data packet with information that allows that packet to be reconstructed in the presence of channel errors. This results in more consistent overall bandwidth in the presence of a noisy channel at the expense of a reduction in the maximum bandwidth achievable; this has been shown to be a very good trade-off for multimedia applications. The TGe proposal is based on well-understood Reed-Solomon coding and supports FEC capability.

- Traffic Prioritization—A newly defined ability is the opportunity to handle traffic differently according to the priority of that traffic. The priority must be specified as part of the traffic and is not determined by the MAC. Also, the scheduling algorithm used to map particular traffic priorities to different channel access methods will probably remain unspecified.

- Traffic Parameterization—Another newly defined ability is the opportunity to handle traffic in accordance with a parameterized specification for each traffic type, allowing for more broadly differentiated traffic than would be available with simple prioritization. The parameterization must be specified as part of the traffic and is not determined by the MAC. Also, the scheduling algorithm used to map particular traffic parameters to different channel access methods will probably remain unspecified.

*Coordination Function Enhancements*—The existing standard defines a Distributed Coordination Function (DCF) and an optional Point Coordination Function (PCF). The new proposals include support for a third type of coordination function with several associated concepts.

- Transmission Opportunity (TxOP)—The existing specification allows only a single data frame to be transmitted in response to a poll; the new proposal includes the concept of a TxOP, which is a duration during which the STA can transmit as many data frames as will fit within the specified duration.

- Hybrid Coordination Function (HCF)—A new coordination function is defined that will allow contention-free access during either the Contention Period (CP) or the Contention-Free Period (CFP). In the CP, this function relies on the concept of a Contention-Free Burst (CFB).

- Contention-Free Burst (CFB)—by taking advantage of the way channel access timing works within the CP, the HCF can contend for the channel and, once access is gained, hold that access for as long as a contiguous string of Polls + TxOPs + acknowledgments can be maintained. CFB map most naturally to the notion of parameterized traffic.

- Reservation Request (RR)—The existing specification does not specify how a STA can be placed on the polling list, so a new RR frame has been defined to allow a STA to be placed on the polling list whenever data are available for transmission, hence making the polling mechanism more efficient.

*Contention Period Enhancements*—The existing IEEE 802.11 specification is divided into a CP and an optional CFP. Specific enhancements have been made to both, but the CP enhancements are discussed here.

- Enhanced DCF (eDCF)—The current proposal includes changes to the Distributed Coordination Function (DCF) that provide for prioritization of traffic at the level of access to the channel. This is achieved by allowing changes to the contention backoff algorithms on a per-priority basis, resulting in a statistical difference in the likelihood of delivering traffic from each of the discrete priorities. This maps very nicely into the diffserv model.

*Contention-Free Period Enhancements*—As indicated, the existing standard includes an optional CFP and an associated PCF.

- HCF in the CFP—Use of the newly proposed HCF is not limited to the CP: it may also be used during the CFP as a replacement for the optional PCF. The advantage in doing so versus using it in the CP is that latencies are likely to be more predictable, because no channel contention is permitted within the CFP.

## 6.5 TGg—2.4-GHz High Rate

Although not exclusively targeted for support of QoS and multimedia applications, any additional bandwidth can certainly be put to good use by those applications. TGg is presently being proposed at rates of approximately twice that available within IEEE 802.11b, or about 20–22 Mbit/s. There are a number of competing proposals, but they all share backward compatibility with the existing 802.11b standard and operation within the 2.4-GHz band. The proposed TGe changes to the MAC will operate with this new high-rate standard.

## 6.6 TGh—Dynamic Channel Selection and Transmit Power Control

Although not specifically targeted toward support of QoS and multimedia applications, the two aspects of this standard lend themselves indirectly to such support.

*Dynamic Channel Selection (DCS)*—Spectrum management is an important although little-recognized aspect of support for QoS and multimedia. Channel error rates in wireless LANs can vary dramatically as a result of stationary and moving obstacles, antenna placement, and in-channel interferers such as microwave ovens, cordless telephones, and wireless personal area networks. The more severe the channel conditions the more likely channel errors will occur, resulting in retransmissions that in turn result in bandwidth and QoS degradation for sensitive applications such as video. DCS provides a mechanism by which an entire wireless LAN Basic Service Set (BSS, or subnet) can be moved to an adjacent channel that may potentially offer better channel conditions, thereby better preserving the QoS characteristics of the network. It also provides a mechanism by which overlapping BSSs from different wireless LANs, such as adjacent apartments, can avoid one another.

*Transmit Power Control (TPC)*—Power management is another little-recognized aspect of QoS support, largely in the context of overlapping BSSs. The TGh-proposed transmit power controls will enable STAs to communicate with each other, and with the AP, at the minimum power required to establish a high-quality link. Although this matters little for the BSS in question, it can matter a great deal to adjacent BSSs in the same channel. Effectively, the interference between adjacent BSSs should be reduced in closely packed environments, thereby elevating the level of QoS for all of the BSSs concerned.

## 6.7 TGi—Security

Security is another aspect of the standards work that does not apply directly to QoS and multimedia. However, two of the very serious concerns expressed by people ·such as cable operators are pertinent to the case of multimedia distribution.

*Stream Theft*—The first concern is the ability for a third party to intercept a link transmission and decode it for its own purposes, thereby gaining access to content that another individual may in fact be paying for, and to gain that access without anyone's knowledge. This issue is best addressed by means of improved encryption.

*Encryption*—The IEEE 802.11 standard specifies an optional form of encryption referred to as Wired Equivalent Privacy (WEP). WEP uses 40-bit keys and the RSA algorithm and is no longer considered very secure. The TGi work includes the proposed addition of two new encryption schemes: WEP2, based on 128-bit keys, and the Advanced Encryption Standard (AES) based on elliptical encryption algorithms. Both of these methods should be considerably more secure from stream theft than the existing WEP.

*Digital Rights Management*—But encryption is only part of the problem. Also a problem is the possibility that a STA may knowingly rebroadcast an encrypted stream to another properly authenticated STA, in effect creating a ministudio. This issue is a problem in the existing 802.11 standard because the process for allocating and distributing encryption keys is not defined.

*Upper Layer Authentication*—Work in TGi includes the concept of key distribution and handling controlled by the layers above the MAC. In this way, the effective owner of the digital rights can determine who is and who is not to be given permission to receive a particular stream.

## 6.8 Conclusions

The convergence of data and entertainment has increased in pace, and there has been a large amount of corresponding activity within the IEEE 802 standards committees to create enhancements to existing standards that will be capable of supporting this convergence. Much of the focus has been in the IEEE 802.11

Wireless LAN committee, and much of the emphasis there has been on explicit support for QoS enhancements and multimedia applications, although other committee activities have proposed enhancements that may also be used to support QoS and multimedia applications. This chapter has described the current state of the proposed changes to the relevant IEEE 802.11 standards and has provided context for the reasons behind those proposals. The indicated conclusion of understanding these proposals and the context in which they are made is that the IEEE 802.11 standard is well positioned to meet a growing need for QoS and multimedia support in wireless LANs.

# Chapter 7

# Overview of Wireless LAN Security

*Cisco Systems*

The free-space wireless link is more susceptible to eavesdropping, fraud, and unauthorized transmission than its wired counterpart. Unauthorized people can tap the radio signal from anywhere within range. If someone sets a mobile client within a wireless coverage area to transmit packets endlessly, all other clients are prevented from transmitting, thus bringing the network down. Being an open medium with no precise bounds makes it impractical to apply physical security as in wired networks. Nevertheless, several security mechanisms can be used to prevent unauthorized access of data transmitted over a wireless LAN.

## 7.1 The Growth of Wireless LANs

Until recently, wireless LAN products were used primarily in certain vertical markets (e.g., retail, education, and health care) where mobile users with a need for LAN access were satisfied with data rates of 2 Mbit/s or less. Even though most wireless LANs were extensions of wired LANs, the proprietary nature and slow speeds of wireless LANs forced organizations to manage wireless LANs as unique entities. To make wireless LANs more "mainstream," customers pressed vendors to develop a high-speed wireless LAN standard that would encourage interoperability, reduce prices, and provide the bandwidth needed by today's business applications. In 1999, the Institute of Electrical and Electronics Engineers (IEEE) ratified an extension to a previous standard. Called IEEE 802.11b, it defines the standard for wireless LAN products that operate at an Ethernet-like data rate of 11 Mbit/s, a speed that makes wireless LAN technology viable in enterprises and other large organizations. Interoperability of wireless LAN products from different vendors is ensured by an independent organization called the Wireless Ethernet Compatibility Alliance (WECA), which brands compliant products as "Wi-Fi." Dozens of vendors market Wi-Fi products, and organizations of every size and type are considering, if not deploying, wireless LANs. Demand for wireless access to LANs is fueled by the growth of mobile computing devices, such as laptops and personal digital assistants, and a desire by users for continual connections to the network without having to "plug in." There will be over a billion mobile devices by 2003, and the wireless LAN market is projected to grow to over US$2 billion by 2002.

## 7.2 The Need for Centralized Security Management

Now that wireless LANs have become mainstream, organizations want to tightly integrate wireless LANs with wired LANs. Network managers are reluctant or unwilling to deploy wireless LANs unless those LANs provide the type of security, manageability, and scalability offered by wired LANs. The chief concern is security, which encompasses access control and privacy. Access control ensures that sensitive data can be accessed only by authorized users. Privacy ensures that transmitted data can be received and understood only by the intended audience. Access to a wired LAN is governed by access to an Ethernet port for that LAN. Therefore, access control for a wired LAN often is viewed in terms of physical access to LAN ports. Similarly, because data transmitted on a wired LAN are directed to a particular destination, privacy cannot be compromised unless someone uses specialized equipment to intercept transmissions on their way to their destination. In short, a security breach on a wired LAN is possible only if the LAN is physically compromised.

With a wireless LAN, transmitted data are broadcast over the air with radio waves, so data can be received by any wireless LAN client in the area served by the data transmitter. Because radio waves travel through ceilings, floors, and walls, transmitted data may reach unintended recipients on different floors and even outside the building of the transmitter. Installing a wireless LAN may seem like putting Ethernet ports everywhere, including in your parking lot. Similarly, data privacy is a genuine concern with wireless LANs because there is no way to direct a wireless LAN transmission to only one recipient. The IEEE 802.11b standard includes components for ensuring access control and privacy, but these components must be deployed on every device in a wireless LAN. An organization with hundreds or thousands of wireless LAN users needs a solid security solution that can be managed effectively from a central point of control. Some cite the lack of centralized security as the primary reason why wireless LAN deployments have been limited to relatively small workgroups and specialized applications.

## 7.3 First-Generation Wireless LAN Security

The IEEE 802.11b standard defines two mechanisms for providing access control and privacy on wireless LANs: Service Set Identifiers (SSIDs) and Wired Equivalent Privacy (WEP). Another mechanism to ensure privacy through encryption is the use of a Virtual Private Network (VPN) that runs transparently over a wireless LAN. Because the use of a VPN is independent of any native wireless LAN security scheme, VPNs are not discussed in this chapter.

### 7.3.1 SSID

One commonly used wireless LAN feature is a naming handle called an SSID, which provides a rudimentary level of access control. An SSID is a common network name for the devices in a wireless LAN subsystem; it serves to logically segment that subsystem. The use of the SSID as a handle to permit/deny access is dangerous because the SSID typically is not well secured. An access point, the device that links wireless clients to the wired LAN, is usually set to broadcast its SSID in its beacons.

### 7.3.2 WEP

The IEEE 802.11b standard stipulates an optional encryption scheme called wired equivalent privacy, or WEP, that offers a mechanism for securing wireless LAN data streams. WEP uses a symmetric scheme in which the same key and algorithm are used for both encryption and decryption of data. The goals of WEP include:

- Access control: Prevent unauthorized users, who lack a correct WEP key, from gaining access to the network;
- Privacy: Protect wireless LAN data streams by encrypting them and allowing decryption only by users with the correct WEP keys.

Although WEP is optional, support for WEP with 40-bit encryption keys is a requirement for Wi-Fi certification by WECA, so WECA members invariably support WEP. Some vendors implement the computationally intense activities of encryption and decryption in software, whereas others, like Cisco Systems, use hardware accelerators to minimize the performance degradation of encrypting and decrypting data streams.

The IEEE 802.11 standard provides two schemes for defining the WEP keys to be used on a wireless LAN. With the first scheme, a set of as many as four default keys are shared by all stations (i.e., clients and access points) in a wireless subsystem. When a client obtains the default keys, that client can communicate securely with all other stations in the subsystem. The problem with default keys is that when they become widely distributed they are more likely to be compromised. In the second scheme, each client establishes a "key mapping" relationship with another station. This is a more secure form of operation because fewer stations have the keys, but distributing such unicast keys becomes more difficult as the number of stations increases.

### 7.3.3 Authentication

A client cannot participate in a wireless LAN until that client is authenticated. The IEEE 802.11b standard defines two types of authentication methods: open and shared key. The authentication method must be set on each client, and the setting should match that of the access point with which the client wants to associate. With open authentication, which is the default, the entire authentication process is done in clear text, and a client can associate with an access point even without supplying the correct WEP key. With shared-key authentication, the access point sends the client a challenge text packet that the client must encrypt with the correct WEP key and return to the access point. If the client has the wrong key or no key, it will fail authentication and will not be allowed to associate with the access point.

Some wireless LAN vendors support authentication based on the physical address, or Medium Access Control (MAC) address, of a client. An access point will allow association by a client only if that client's MAC address matches an address in an authentication table used by the access point.

## 7.4 Security Threats

### 7.4.1 Theft of Hardware

It is common to statically assign a WEP key to a client, either on the client's disk storage or in the memory of the client's wireless LAN adapter. When this is done, the possessor of a client has possession of the client's MAC address and WEP key and can use those components to gain access to the wireless LAN. If multiple users share a client, then those users effectively share the MAC address and WEP key. When a client is lost or stolen, the intended user or users of the client no longer have access to the MAC address or WEP key and an unintended user does. It is next to impossible for an administrator to detect the security breach; a proper owner must inform the administrator. When informed, an administrator must change the security scheme to render the MAC address and WEP key useless for wireless LAN access and decryption of transmitted data. The administrator must recode static encryption keys on all clients that use the same keys as the lost or stolen client. The greater the number of clients, the larger the task of reprogramming WEP keys. What is needed is a security scheme that:

- Bases wireless LAN authentication on device-independent items such as usernames and passwords, which users possess and use regardless of the clients on which they operate;
- Uses WEP keys that are generated dynamically on user authentication, not static keys that are physically associated with a client.

### 7.4.2 Rogue Access Points

The 802.11b shared-key authentication scheme employs one-way, not mutual, authentication. An access point authenticates a user, but a user does not and cannot authenticate an access point. If a rogue access point is placed on a wireless LAN, it can be a launch pad for denial-of-service attacks through the "hijacking" of the clients of legitimate users. What is needed is mutual authentication between the client and an authentication server whereby both sides prove their legitimacy within a reasonable time. Because a client and an authentication server communicate through an access point, the access point must support the mutual authentication scheme. Mutual authentication makes it possible to detect and isolate rogue access points.

### 7.4.3 Other Threats

Standard WEP supports per-packet encryption but not per-packet authentication. A hacker can reconstruct a data stream from responses to a known data packet. The hacker then can spoof packets. One way to mitigate this security weakness is to ensure that WEP keys are changed frequently. By monitoring the 802.11 control and data channels, a hacker can obtain information such as:

- Client and access point MAC addresses;
- MAC addresses of internal hosts;

- Time of association/disassociation.

The hacker can use such information to do long-term traffic profiling and analysis that may provide user or device details. To mitigate such hacker activities, a site should use per-session WEP keys.

### 7.4.4 Addressing Security Threats

In summary, to address the security concerns raised in this section, a wireless LAN security scheme should:

- Base wireless LAN authentication on device-independent items such as usernames and passwords, which users possess and use regardless of the clients on which they operate;
- Support mutual authentication between a client and an authentication (RADIUS) server;
- Use WEP keys that are generated dynamically on user authentication, not static keys that are physically associated with a client;
- Support session-based WEP keys.

First-generation wireless LAN security, which relies on static WEP keys for access control and privacy, cannot address these requirements.

## 7.5 A Complete Security Solution

What is needed is a wireless LAN security solution that uses a standards-based and open architecture to take full advantage of 802.11b security elements, provide the strongest level of security available, and ensure effective security management from a central point of control. A promising security solution implements key elements of a proposal jointly submitted to the IEEE by Cisco Systems, Microsoft, and other organizations. Central to this proposal are the following elements:

- Extensible Authentication Protocol (EAP), an extension to Remote Access Dial-In User Service (RADIUS) that can enable wireless client adapters to communicate with RADIUS servers;
- IEEE 802.1x, a proposed standard for controlled port access.

When the security solution is in place, a wireless client that associates with an access point cannot gain access to the network until the user performs a network logon. When the user enters a username and password into a network logon dialog box or its equivalent, the client and a RADIUS server (or other authentication server) perform mutual authentication, with the client authenticated by the supplied username and password. The RADIUS server and client then derive a client-specific WEP key to be used by the client for the current logon session. All sensitive information, such as the password, is protected from passive monitoring and other methods of attack. Nothing is transmitted over the air in the clear.

As illustrated in Figure 7.1, the sequence of events follows:

- A wireless client associates with an access point;
- The access point blocks all attempts by the client to gain access to network resources until the client logs on to the network;
- The user on the client supplies a username and password in a network logon dialog box or its equivalent;
- Using 802.1x and EAP, the wireless client and a RADIUS server on the wired LAN perform a mutual authentication through the access point. One of several authentication methods or types can be used. With the Cisco authentication type, the RADIUS server sends an authentication challenge to the client. The client uses a one-way hash of the user-supplied password to fashion a response to the challenge and sends that response to the RADIUS server. Using information from its user database, the RADIUS server creates its own response and compares that to the response from the client. Once the RADIUS server authenticates the client, the process repeats in reverse, enabling the client to authenticate the RADIUS server;
- When mutual authentication is successfully completed, the RADIUS server and the client determine a WEP key that is distinct to the client and provides the client with the appropriate level of network access, thereby approximating the level of security inherent in a wired switched segment to the individual desktop. The client loads this key and prepares to use it for the logon session;
- The RADIUS server sends the WEP key, called a session key, over the wired LAN to the access point;
- The access point encrypts its broadcast key with the session key and sends the encrypted key to the client, which uses the session key to decrypt it;
- The client and access point activate WEP and use the session and broadcast WEP keys for all communications during the remainder of the session.

Support for EAP and 802.1x delivers on the promise of WEP, providing a centrally managed, standards-based, and open approach that addresses the limitations of standard 802.11 security. In addition, the EAP framework is extensible to wired networks, enabling an enterprise to use a single security architecture for every access method.

## 7.6 Conclusions

It is likely that dozens of vendors will implement support for 802.1x and EAP in their wireless LAN products. Knowing the customer benefits of 802.1x, Cisco Systems supports the forthcoming standard today, offering a complete, end-to-end security solution that is fully compliant with 802.1x. With the Cisco Systems wireless LAN security solution in place, an organization:

- Minimizes the security threats of lost or stolen hardware, rogue access points, and hacker attacks;

- Uses user-specific, session-based WEP keys created dynamically at user logon, not static WEP keys stored on client devices and access points;
- Manages the security for all wireless users from a central point of control.

Cisco wireless security services closely parallel security available in a wired LAN, fulfilling the need for a consistent, reliable, and secure mobile networking solution.

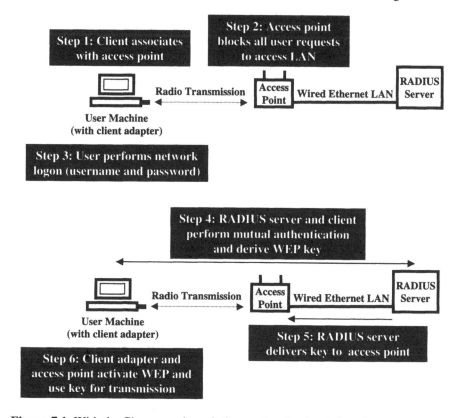

**Figure 7.1.** With the Cisco security solution, authentication is based on username and password, and each user gets a unique, session-based encryption key.

# Chapter 8

# Wireless Network Security

Dorothy Stanley
*Agere Systems, Inc.*[1]

Much attention has been focused recently on the security aspects of existing 802.11b wireless LAN systems. With the rapid growth and deployment of these systems a wide range of networks and applications comes the need to support security solutions that meet the needs of a wide variety of customers. This chapter discusses the topics of encryption, key management, and end user authentication, beginning with current issues and available solutions. Example applications of security solutions are given, using existing Agere Systems wireless LAN products.

## 8.1 Introduction

The level of required security changes over time, as technology and export regulations change and as the processing capabilities of both valid users and potential attackers increase. One static aspect, however, is the need for end users to adhere to recommended security practices, such as keeping up-to-date virus software and intrusion detection software on their laptops. In addition, there are conflicting requirements of security and convenience. End users may desire a simple logon using stored passwords on the laptops, but strong authentication [1] requires two separate credentials from the user. A familiar example of strong authentication can be found in the banking industry: To withdraw cash from an ATM, you are required to present both a password (something you know) and the appropriate ATM card (something you have).

The level of privacy and authentication required may also be a function of the application or location in which the wireless LAN is deployed. Enterprise applications may have different needs than public space applications. A given residential application may need the security level of an enterprise. The security solutions must be broad enough to support this variety of application spaces. The solutions must be easy to use, because the same laptops and devices will be used for Internet access in all of these locations.

---

[1] ORiNOCO is a trademark of Agere Systems. All other trademarks are the property of their respective owners.

## 8.2 Current Issues and Application Solutions

Recently, much attention has been paid to the fact that Wired Equivalent Privacy (WEP) encryption defined by IEEE 802.11 is not an "industrial-strength" encryption protocol. Papers by Borisov [2] and Walker [3] have discussed the vulnerabilities of WEP. The results of Fluhrer et al. [4] have enabled easy-to-mount passive attacks [5] on WEP. This section presents an overview of solutions available today, prior to the availability of an effective 802.11 security standard, that prevent or avoid the identified flaws in WEP.

One way to address the MAC level WEP encryption flaws is to deploy encryption above the MAC layer. The IEEE and the Wireless Ethernet Compatibility Alliance (WECA) advise, "Security layers can be deployed above the wireless LAN layer. An example here is the use of Virtual Private Networks (VPNs), which provides end-to-end security, for which the wireless LAN is transparent" [6]. VPN overlays can be costly, requiring additional equipment (VPN servers) and maintenance expenses. However, some enterprise applications can be well served by the VPN solution. IPSec clients are typically deployed for remote access. The same VPN infrastructure can be used for wireless access clients. Most wireless LAN access points, including the ORiNOCO Access Points AP-500, AP-1000, and AP-2000 are designed to work transparently with VPN solutions.

For some customers, the deployment of low-cost access points coupled with application-level security solutions is desirable. In public space deployments, for example, many operators' primary concern is with economically deploying wireless access networks. Service providers (e.g., Wayport) offer Wi-Fi[2] interoperable access to the greatest number of customers and provide easy-to-use Web portal interfaces for customer registration. Security is provided at the network level, via IPSec VPN access to corporate networks, and for example, HTTPS web server access.

Applications that have deployed an 802.1x [7] authentication infrastructure can also use the automatic rekeying capabilities provided by the 802.1x Port-Based Network Access Control standard, to more frequently change the encryption key that is used. This approach does not correct the underlying flaws in WEP. It minimizes the risk of a passive attack with a tool such as Airsnort, by changing the keys before the required number of packets needed to mount the attack can be collected. Access points such as the ORiNOCO AP-2000 support this method as well as weak-key avoidance, which defeats the specific Airsnort attack.

For applications requiring stronger encryption at the wireless LAN MAC layer, where VPNs are not used, proprietary solutions are available. For example, the ORiNOCO Access Server (AS-2000) provides non-WEP encryption in software, which avoids many of the vulnerabilities identified in WEP, and is session based rather than frame (packet) based. This implementation does not reinitialize the algorithm state for each packet as is done for WEP. Instead, the algorithm state from the end of one packet is used to begin encryption with the next packet. In addition, unique per-user keys are used for encryption of traffic to and from each station. The per-user session-based encryption supported by the Access Server provides protection against passive eavesdropping attacks. The Reefedge connect server [8] and the Musenki layer 2 IPSec [9] are examples of other proprietary solutions.

---

[2] Wi-Fi is a trademark of the Wireless Ethernet Compatibility Alliance (WECA).

## 8.3 MAC-Level Encryption Enhancements

In addition to the security solutions described above, MAC-level encryption enhancements are being specified to provide standard, improved encryption and data authentication at the wireless MAC level and to standardize use of upper-layer authentication. The 802.11 Security Subgroup, 802.11i [10], is developing these standards for future use in 802.11 networks. This section describes two possible improved encryption methods:

• A strengthened version of the RC-4/per-frame IV encryption algorithm;
• A 128-bit AES encryption algorithm.

Improvements and enhancements that address the shortcomings of WEP have been identified on the basis of feedback from members of the cryptographic community. These enhancements include:

• The addition of a per-packet hash function and IV sequencing rules [11, 12];
• The addition of temporal key derivation algorithms [13];
• The addition of rekey mechanisms [13] (described in Section 8.4);
• The addition of a message authentication code, termed message integrity code;

Taken together, the enhanced protocol, known as the Temporal Key Integrity Protocol (TKIP), addresses the flaws identified in the current WEP algorithm. A critical constraint placed on the strengthened WEP algorithm definition is that it must be able to be implemented and deployed via software upgrade to the existing base of millions of 802.11 devices.

### 8.3.1 A Per-Packet Hash Function

The RC4 key used to encrypt a given data frame in 802.11 WEP is a combination of the initialization vector (IV) and the secret key. Unfortunately, in the key-scheduling algorithm of RC4, the first bytes of the key stream are predictable for certain known IV values [4]. Because the IV used to encrypt a given frame is sent in the clear, a passive observer can easily identify the frames to target for attack. The per-packet hash function is introduced primarily to eliminate this flaw in WEP. The hash function is also defined to include the MAC address of the transmitting station in the hash function. This enables each transmitting station to generate a unique IV stream and thus prevents the reuse of IV values among stations using a shared secret key. IV values must not be reused, to prevent the reuse of RC4 key streams and subsequent data recovery attacks.

A simplified description of a per-packet hash algorithm is shown below. The details of the hash function are provided in Reference 11. The algorithm is described in two phases, both of which use S-boxes to mix and substitute 16-bit values. In phase 1, the 128-bit temporal key and the high 32 bits of the transmitting station's MAC address are hashed into a 128-bit value, composed of 8- to 16-bit values, as illustrated in Figure 8.1.

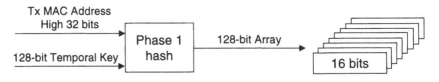

**Figure 8.1.** Phase 1 hash.

Phase 2 of the temporal key hash function takes the 128-bit array from phase 1, together with the IV, and generates a 128-bit per-packet key. As the name implies, the key that is generated will be used for one packet only; the phase 2 hash is calculated for each packet, which is encrypted (Fig. 8.2). The per-packet key is subsequently used as a WEP key, with the first 24 bits transmitted in the clear.

The phase 2 hash consists of three steps: an S-box mixing function that operates on 16-bit values of the array, a mixing function that uses rotate and addition operations, and calculation of the 24-bit WEP IV value. The phase 2 hash eliminates the effects of the WEP/RC4 key scheduling algorithm flaw.

Modifications to the per-packet hash function described above are defined for use in an extended IV, such as a 48-bit IV [12]. In phase 1 of the extended IV, portions of the temporal key, the transmitting station's MAC address, and the IV are hashed into an 80-bit value. Phase 2 of the temporal key hash function takes the 80-bit array from phase 1, together with portions of the IV and the temporal key, to generate a 128-bit per-packet key. Use of an extended IV is attractive, since it virtually eliminates the need to rekey due to exhaustion of the IV space.

### 8.3.2 A Temporal Key Derivation Method

Temporal key derivation defines a method whereby the "secret key" or master key is not used to encrypt data packets but rather is the basis from which temporal or transient encryption keys are derived (Fig. 8.3). These temporal keys may then be used as input to the per-packet hash function described above. Note that this approach is very different from the initial 802.11b definition and implementations, in which the provisioned key is used directly as the secret portion of the encryption key.

One proposed approach [13] for deriving the base and temporal keys is to use a pseudorandom function (PRF), operating on the secret key, a text string, the MAC address of one of the stations, and a nonce value, to generate a temporal key string. The keying material is then used for the encryption and MIC temporal keys (described in Section 8.3.3). Figure 8.4 shows how a group secret key and the PRF are used to derive the transient key material, including a temporal key. The transient key provides the key material for the TKIP per-packet RC4 encryption and the Michael authentication function.

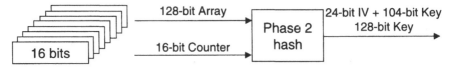

**Figure 8.2.** Phase 2 hash.

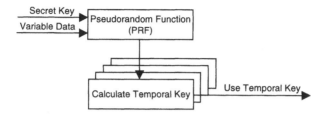

**Figure 8.3.** Temporal key derivation concept.

The use of temporal keys eliminates the issue of IV reuse in WEP because of the relatively small 24-bit IV space associated with using a single secret key for encryption. A new temporal key must be used before the exhaustion of the IV space for the temporal key being used for encryption. Thus the size of the IV space determines how often a new temporal key must be calculated. The per-packet hash function described in Figures 8.1 and 8.2 above supported a 16-bit IV, requiring a new temporal key every $2^{16}$, or 65K packets. This is a small number of packets. Therefore, an extended IV, such as the 48-bit IV space is attractive; quadrillions of packets can be sent before the IV space $2^{48}$ is exhausted. Local policy will require rekeying prior to extended IV exhaustion. For example a local policy may require new broadcast keys every day, so that employees who have left the company cannot find a clever way to recover a key from their system and use it to passively read and decrypt broadcast traffic.

It is critical that for any encryption key, a given IV be used to encrypt one and only one packet. Proper use of IVs is ensured by the application of IV sequencing rules. First, the notion of a sequence counter is introduced. The sequence counter is incremented on a per-frame basis. As part of the per-packet mixing function, the sequence counter is mapped to the WEP IV. In addition, the receiver must verify that the IV of a received frame increment sequentially, per quality of service traffic class. If the IV is not incremented properly, the frame is discarded by the receiver.

### 8.3.3 Message Integrity Code

A message integrity code (MIC)[3] is needed to verify the authenticity of the transferred data packet. Use of the MIC verifies that the packet was not modified in transit and that the source and destination addresses were not changed. The ability to verify message integrity is viewed by cryptographers to be as important as, if not more important than, the privacy provided by encryption. The MIC is required to prevent the "bit-flipping" attacks identified in Reference 2. A MIC algorithm known as "Michael" is proposed to be the TKIP MIC [14].

### 8.3.4 AES Encryption—Beyond RC4 and WEP

Encryption at the MAC level can also be strengthened by proper use of additional encryption algorithms. The Advanced Encryption Standard (AES) Rijndael algorithm [15] has been selected by NIST [16] as the next-generation encryption

---

[3] Here the term "MIC" is used, as "MAC" is already used for Medium Access Control. Message Authentication Code (MAC) is the standard cryptographic term.

algorithm, to replace DES and 3DES. Several modes or ways of using the AES algorithm have been defined. Two of these, AES-OCB [17] and AES-CBC-MAC [18] are of particular interest for wireless LAN application. AES-OCB mode has the attribute of providing both authentication and encryption in one pass through the data. There are some concerns over intellectual property claims, however. AES-CBC-MAC, which combines Counter mode encryption with Cipher Block Chaining message integrity/authentication is termed AES-CCM [19]. It is expected that one mode of AES encryption will be defined for IEEE 802.11 encryption in the enhanced security standard, IEEE 802.11i.

## 8.4 Encryption Key Definition and Distribution

This section describes the current 802.11b key definitions [20] and presents alternatives for periodic rekeying of encryption keys. The 802.11 standard defines two methods for using WEP keys between two stations: default keys (DK) and mapped keys (MK). Each of these is summarized below.

### 8.4.1 Definition of Default Keys

The default key method is mandatory for Wi-Fi and is supported by most product implementations today. It is based on the use of a set of four default keys in each station, one of which is designated as a transmit key. All four keys (if configured) can be used to decrypt incoming frames. The IV field in the received frame indicates which default key was used for encryption, via two KeyID bits. This same key must be used for decryption in the receiving station.

Default keys can be used to support stations using shared keys (all stations and AP use the same key) and for stations using individual station keys. Default keys are used in all products today supporting shared keys. When default keys are used to support individual station keys, the AP must keep a copy of the default keys on a per-MAC-address basis.

As shown in Figure 8.5, the "table" contains up to four key values. The "ID" contains the key-offset in the table to use for transmitting frames. Note that the reference to the default keys (1–4) is one higher than the associated ID (0–3).

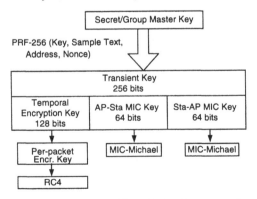

**Figure 8.4.** Temporal key derivation example.

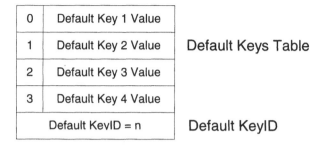

| 0 | Default Key 1 Value | **Default Keys Table** |
|---|---|---|
| 1 | Default Key 2 Value | |
| 2 | Default Key 3 Value | |
| 3 | Default Key 4 Value | |
| Default KevID = n | | **Default KeyID** |

**Figure 8.5.** Default keys table and default KeyID.

### 8.4.2 Definition of Mapped Keys

The mapped key method provides one way to use a unique key for a connection between two stations. A station using this method keeps a table that maps MAC addresses to key values (Fig. 8.6). When a frame is to be transmitted to a station that has a MAC address in the table, the mapped key is used to encrypt the frame. When a frame is received from a station that has a MAC address in the table, the mapped key is used to decrypt. Frames sent this way must set the KeyID bits in the IV field of the frame to zero (00). Operating with this method means that the tables in two stations that need to communicate must contain each other's MAC address and must map these MAC addresses to the same key value.

An AP can support both default keys and mapped keys simultaneously. The standard specifies that the key mappings method must be used if a key is present. Default keys must be used only when no entry is present for key mappings. Techniques for rekeying are now described, which use both the default key and mapped key capabilities.

### 8.4.3 Rekeying Methods Using Default Keys and Mapped Keys

This section describes three approaches that can be used for rekeying an encryption key in 802.11 systems:

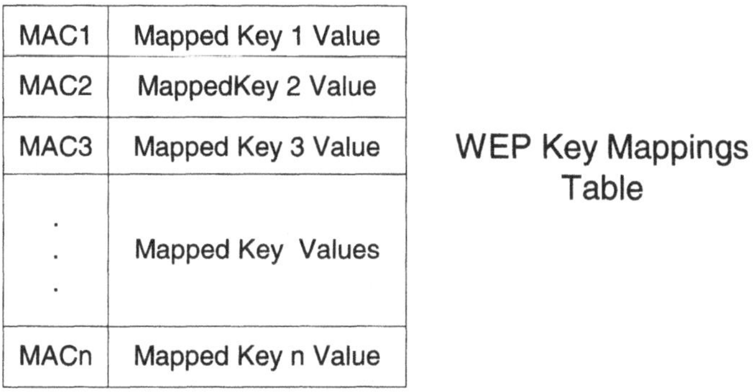

| MAC1 | Mapped Key 1 Value | **WEP Key Mappings Table** |
|---|---|---|
| MAC2 | MappedKey 2 Value | |
| MAC3 | Mapped Key 3 Value | |
| . . . | Mapped Key Values | |
| MACn | Mapped Key n Value | |

**Figure 8.6.** Mapped keys table.

- Use of Implicit Time-Based Methods;
- Use of MAC-Level Management Messages;
- Use of Upper-Layer Authentication Messages, 802.1x EAPOL-Key. This is the approach adopted in TKIP.

### 8.4.3.1 Time-Based Methods

Time-based methods use elements in the 802.11 beacon messages to provide data for temporal key generation and to indicate when the key is to change. Two examples are described here, use of the existing Targeted Beacon Transmit Time (TBTT) [21], and addition of a rekey information element in the beacon [22].

The state of the TBTT (bit-19 toggles every second) can be used to determine whether a new secure hash is needed and which key set is to be used (Fig. 8.7). This method provides a way to, for example, generate and use a new key every second. The default keys can be used, enabling a simple transition from one key to the next. This method works well for default keys shared among stations. When it is used with default or mapped session keys, the AP must change keys for multiple stations simultaneously. This is difficult to support in practice, when the AP serves a large number of stations, because the new keys for each station must be calculated at the same time. With the TBTT method, the sets of default keys change over time so that at any moment in time there is the correct overlap between all involved stations to ensure that a receiving station uses the proper key to decrypt the received packet.

A second alternative is to modify the beacon message to include the information needed to securely synchronize the encryption keys. A new rekey information element can be used for this, and the contents of one possible rekey information element are shown in Figure 8.8 [22].

- The Nonce is an 8-byte nonce;
- The Cipher Suite is the cipher suite enforcing this rekey protocol;
- The KeyID indicates the agreed upon index used for this rekeying;
- The Key Sequence Value is the index of the temporal encryption key to activate;
- The Rekey Count indicates the number of beacons (including the current beacon frame) appear before the next rekey will occur;
- The Rekey Period indicates the number of beacon intervals;
- The MIC is the frame's message integrity check. The message integrity check is computed over the Rekey Information Element.

The advantage of using a new element in the beacon is that a MIC over the beacon element is provided, preventing beacon elements from being forged by an attacker.

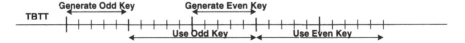

**Figure 8.7.** Key change with TBTT.

| 8 octets | 4 octets | 1 octet | 4 octets | 1 octet | 1 octet | 8 octets |
|----------|----------|---------|----------|---------|---------|----------|
| Nonce | Cipher Suite | KeyID | Key Sequence Number | Rekey Count | Rekey Period | MIC |

**Figure 8.8.** Example beacon re-key information element.

### 8.4.3.2 Use of MAC-Level Management Messages

In this approach, the indication to initialize keys and rekey to new temporal keys is explicitly provided through the introduction of new MAC-level management messages. Five new messages are defined: Enable request, Enable response, Transition request, Transition response, and Transition confirm.

A simple example is shown in Figure 8.9, with a detailed description provided in Reference 22. This approach is required for applications that use the existing key mapping keys for individual station keys. The advantage of this approach is that the transition to a new key is unambiguous. Both sides verify that they have no additional messages to send that could be encrypted under a previous key.

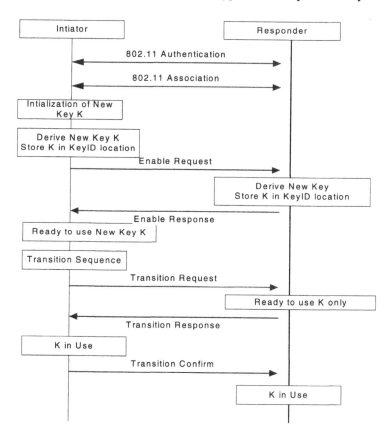

**Figure 8.9.** An example of rekey message flow.

### 8.4.3.3 Use of Upper-Layer 802.1x EAPOL-Key Authentication Messages

Another alternative to accomplish rekeying is to use the currently defined 802.1x EAPOL-Key message, implementing re-keying either at the upper layer (where it is defined) or at the MAC layer. The EAPOL Key message is being used today in 802.1x EAP-TLS (Transport Level Security) [23] implementations, for example, to rekey both shared key (default key) and per-station key (mapped key) systems. This message is always sent from the AP to the stations and is transported as an 802.1x control protocol message. One drawback to this approach is that no acknowledgment is sent to indicate that the station is ready to receive packets. If this message is received and acknowledged (by the 802.11 layer) but incorrectly processed in the application layer, then the encryption between the AP and station will not be synchronized properly. This is an infrequent failure case.

New EAPOL-Key message types are proposed and defined in [13]. An AES-based key transport is defined, together with a four-message protocol exchange, which indicate that keys are established and available for use.

### 8.4.4 Secret Key Distribution and Generation

In the existing 802.11b specification, key distribution mechanisms are not defined. The default key and mapped key structures assume that the keys are provided via an upper layer or manual method. Automated key distribution via custom script tools and manual key entry are available and in use today.

Upper-layer authentication methods can be used to distribute the encryption keys. Many access points, including the AP-2000, support 802.1x EAP-TLS authentication. The EAP-TLS protocol generates secret keys at the station and the RADIUS server. The RADIUS server delivers the secret key to the AP, eliminating the need to manually distribute a WEP key. This secret or session key can be used for encryption directly or to encrypt the WEP keys as they are delivered to the stations. The AP-2000 will also automatically rekey a station once the station has been authenticated. Frequent rekeying mitigates one major attack on WEP while avoiding additional loading of the authentication servers for frequent reauthentication.

Automatic key generation is supported in the ORiNOCO Access Server-2000, using the Diffie–Hellman algorithm between the station and the AS-2000, eliminating the need to configure or distribute WEP keys. The implementation of the algorithm uses a 768-bit strong prime and a strong random number generator to provide a secure implementation. Unique keys are generated and used per-station, in both the direction from the AS-2000 to the station and the direction from the station to the AS-2000.

## 8.5 Authentication

Authentication of end users or end systems is needed to control access to the local LAN. In enterprise applications, only authorized users must be allowed to access the corporate intranet. In public space applications, user identification is needed by the service provider to accurately bill the end user. This section gives a brief overview of six authentication methods: EAP-TLS, EAP-MD5, EAP-TTLS, Access Server Authentication, and, for completeness, 802.11 and RADIUS MAC-based authentication.

There are two major ways to authenticate an end user or device: digital certificates and shared secrets (passwords). The most common certificate-based authentication method is EAP-TLS. Wireless LAN access point products supporting EAP-TLS became available in late 2001. Although each authentication method has advantages and disadvantages [24], the needs of individual deployments may require use of a method supporting a specific type of user credential.

### 8.5.1 802.1x EAP Authentication

The EAP-TLS [23] protocol provides a mechanism for certificate-based mutual authentication, together with the establishment of a secret key at the station and the RADIUS server, and AP. It requires prior distribution of client side and server side certificates via a secure wired connection to the target network. RADIUS Authentication servers supporting EAP-TLS, and certificate management capabilities are also required. A simplified message diagram for EAP-TLS is shown in Figure 8.10. EAP authentication messages sent to/from the station to the RADIUS Server transit the AP.

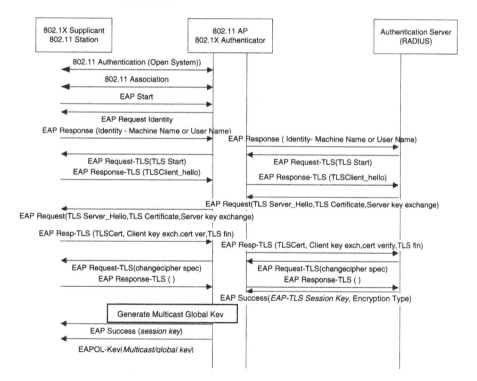

**Figure 8.10.** Simplified EAP-TLS message flow.

### 8.5.2 EAP-MD5

The EAP-MD5 [25] authentication algorithm provides one-way password-based network authentication of the client. It is expected to be widely used in 802.1x wired Ethernet switch deployments. This algorithm can also be used for wireless applications with less stringent wireless LAN security requirements. For example, use of EAP-MD5 authentication may be sufficient for a public space application, in which encryption is provided at the application level. The disadvantage of using EAP-MD5 in wireless LAN applications is that no encryption keys are generated. Also, although the protocol can be used by the client to authenticate the network, it is typically used only for the network to authenticate the client. Finally, as the disassociate message is not authenticated, a valid established session can be hijacked by an attacker [26]. The message flow is shown in Figure 8.11.

### 8.5.3 EAP-TTLS

EAP-TTLS [20] can be viewed as an interesting combination of both EAP-TLS and traditional password-based methods such as Challenge Handshake Authentication Protocol (CHAP) [21], and One Time Password (OTP). In this method, a TLS tunnel is first established between the station supplicant and the authentication server. The client authenticates the network to which it is connecting by authenticating the digital certificate provided by the TTLS server. This is exactly analogous to the techniques used to connect to a secure web server. Once an authenticated "tunnel" is established, the authentication of the end user occurs. EAP-TTLS has the added benefit of protecting the identity of the end user from view over the wireless medium. In this way anonymity of the end user, a desirable attribute, is provided. EAP-TTLS also enables existing end user authentication systems to be reused. The simplified message protocol exchange for EAP-TTLS is shown in Fig. 8.12.

The benefit of the using EAP for identification of the authentication types is that additional EAP types can be easily defined and added to a system. EAP types being specified include EAP-SIM [29], which reuses the mobile GSM authentication credentials, EAP-SRP [30], a secure password-based method, and EAP-SKE [31], which reuses the mobile IP cryptographic secret credentials. EAP-PEAP [32] is similar to EAP-TTLS in concept, but tunnels only EAP-based authentication.

The discussion above focused on standard and published draft EAP methods. Proprietary EAP authentication methods also exist, most notably Cisco's LEAP, a password-based solution [33].

## 8.6 Access Server Authentication

This section describes the authentication methods supported by the ORiNOCO Access Server. The overall approach is similar to that of EAP-TTLS but using a Point-to-Point (PPP) rather than EAPOL implementation. A secure tunnel is first established between the station and the AS-2000. This is followed by a standard PPP Password Authentication Protocol (PAP) or CHAP authentication.

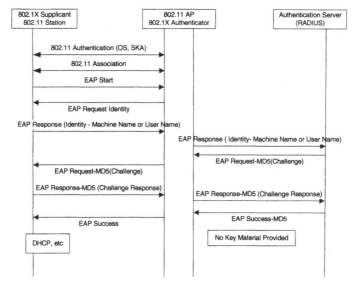

**Figure 8.11.** EAP-MD5 message flow.

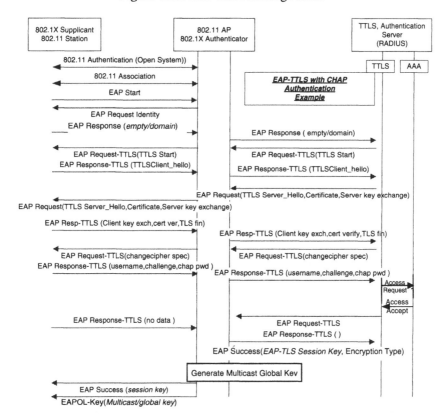

**Figure 8.12.** Simplified EAP-TTLS message flow.

### 8.6.1 Access Server Authentication Using CHAP

The end station is configured to require use of PPP CHAP for authentication. The message flow for authentication using CHAP is presented in Figure 8.13. The client requests CHAP during the PPP link establishment phase. The PPP server in the AS-2000 Access Server responds with a CHAP challenge consisting of a challenge ID and a challenge value, which is a randomly generated number. The client formulates a challenge response consisting of its username and output from a Message Digest 5 (MD5) forward-hashing algorithm [25]. The MD5 algorithm takes as input a message of arbitrary length and produces as output a 128-bit "fingerprint" or "message digest." Input to the MD5 algorithm includes the challenge ID, shared secret (password), and challenge value generated by the PPP server.

The challenge response received by the PPP server is encapsulated in a RADIUS Access Request message and sent to the RADIUS server. An attacker who has obtained a username and hashed password would not be able to replay this information to an AS-2000 Access Server in the target network because the Access Server would generate a new CHAP challenge value. An attacker could retrieve the username but would have to reverse the MD5 forward-hashing function to extract the user password. Policies that require the use of strong, nondictionary passwords are used to prevent potential dictionary attacks.

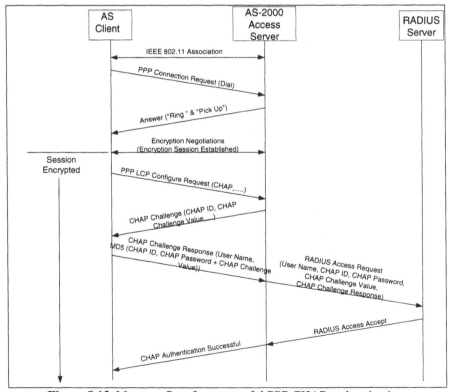

**Figure 8.13.** Message flow for successful PPP CHAP authentication.

### 8.6.2 Access Server Authentication Using PAP

Some PAP-based authentication systems remain in use today, frequently associated with proxy-RADIUS hierarchies of authentication servers. With PAP authentication, the user's password is processed with an MD5 hash at the RADIUS client in the access server rather than at the end station. The end station sends the clear text password to the access server, and this exchange is protected from external sniffing by the encrypted Diffie–Hellman tunnel established between the station and the access server. The access server hashes the password and encrypts this hash with the RADIUS shared secret before sending it to the RADIUS server. These systems are vulnerable to man-in-the-middle attacks only by an attacker who either successfully attacks the MD5-hashed password recovered at the RADIUS server or creates his/her own access server software load to obtain the password before it is hashed. Both of these attacks are extremely difficult to accomplish.

### 8.6.3 Access Server Authentication Using Secure Token to Provide "One Time Password"

IT management can configure the network RADIUS server to use a secure token such as RSA's SecurID[®4] to provide a "one time password" for network access for both dial-in and wireless users. Attackers would have to either capture and use the password within less than one minute (time between password changes) or steal a secure token and the user's password to gain access to the network. All common RADIUS servers can be configured to work with SecurID.

## 8.7 IEEE 802.11 and RADIUS MAC Authentication

The IEEE 802.11 standard supports two subtypes of MAC layer authentication services: open system and shared key. Open system authentication is the default authentication service that announces the desire of a station to associate with another station or access point and is used in ORiNOCO products. 802.11 open system authentication at the MAC level is also used with standard 802.1x EAP implementations. 802.11 shared key authentication is used by some systems, and it provides the ability to verify that the AP and the station share the same WEP key before 802.11 associations. A challenge-response protocol is used, and vulnerabilities have been identified. Shared key authentication is not included in the Wi-Fi compliance requirements and is not recommended for use by Wi-Fi.

RADIUS-based MAC authentication is another technique supported by most APs. The MAC addresses of valid 802.11 devices are provisioned into the AP, and only traffic from these MAC addresses will be allowed through the AP. Authentication is tied to the hardware that a person is using and not to the identity of the user. Software does exist to change the MAC address of a wireless device, and thus MAC-based authentication provides a minimal level of access control to wireless networks.

---

[4] Trademark of RSA Security, Inc.

## 8.8 Evolution, Standards, and Industry Efforts

The future growth of 802.11 wireless LANs requires interoperable, standards-based, evolvable solutions that extend the features and security capabilities of 802.11b systems and support the bandwidth needs of 802.11g and 802.11a systems as well. Improved encryption, authentication, and key management capabilities are essential to this effort. Support for the 802.11i standard is expected to be included in most vendors' products. Agere Systems ORiNOCO products today support a range of security capabilities, matching the range of deployment and application needs of the wireless LAN customer base. These capabilities are being continually enhanced through the support of additional encryption, key exchange, and authentication methods.

## Acknowledgments

In the hectic days of writing this chapter, I had helpful suggestions and comments from three people: Onno Letanche, Leo Monteban, and Laura Stanley.

## References

[1]  M. Lobel, "The Case for Strong Authentication," 1999, http://www.pwcglobal.com/extweb/manissue.nsf/DocID/728D168E9E5CCE0 4852566FD00665839

[2]  N. Borisov, I. Goldberg, and D. Wagner, "802.11 Security," http://www.isaac.cs.berkeley.edu/isaac/wep-faq.html

[3]  J. Walker, "Unsafe at Any Key Size: An Analysis of the WEP Encapsulation," November 2000.

[4]  S. Fluhrer, S. Mantin, and A. Shamir, "Weaknesses in the Key Scheduling Algorithm of RC4," Eighth Annual Workshop on Selected Areas in Cryptography (August 2001).

[5]  http://sourceforge.net/projects/airsnort

[6]  IEEE Wi-Fi WEP Security, www.wi-fi.org/pdf/Wi-FiWEPSecurity.PDF

[7]  Standards for Local and Metropolitan Area Networks: Port-Based Network Access Control, International Standard ISO/IEC, IEEE P802.1x, April 2001.

[8]  See http://www.reefedge.com

[9]  See http://www.musenki.com

[10]  http://grouper.ieee.org/groups/802/11

[11] R. Housley and D. Whiting, "Temporal Key Hash," IEEE 802.11-01/550.

[12] R. Housley, D. Whiting, and N. Ferguson, "Alternative Temporal Key Hash," IEEE 802.11i, 11-02-282r0.

[13] T. Moore and C. Chaplin, "TGi Security Overview," IEEE 802.11i, 11-02-114r1.

[14] N. Ferguson, "Michael-an-improved-MIC-for-802.11-WEP," IEEE 802.11i, 11-02-020r0.

[15] http://www.esat.kuleuven.ac.be/~rijmen/rijndael

[16] http://www.nist.gov

[17] P. Rogaway, http://www.cs.ucdavis.edu/~rogaway

[18] See work in progress—S. Frankel, et al., "The AES Cipher Algorithm and Its Use With IPSec," http://www.ietf.org/internet-drafts/draft-ietf-ipsec-ciph-aes-cbc-02.txt

[19] R. Housley, D. Whiting, and N. Ferguson, "AES-CTR-Mode-with-CBC-MAC," 80211-02-001r1.

[20] IEEE 802.11b, See section 8.3.2 (pp. 65–69) and annex D (pp. 477–481).

[21] W. Diepstraten, et al., "Extended WEP Proposal," August 2001, draft.

[22] N. Cam-Winget, et al., "Authenticated Key Exchange at the MAC Layer," IEEE 802.11-01/508r1.

[23] B. Aboba and D. Simon, "PPP EAP TLS Authentication Protocol," IETF RFC 2716, http://www.ietf.org/rfc/rfc2716.txt

[24] C. Ellison and B. Schneier, "Ten Risks of PKI: What You're Not Being Told About Public Key Infrastructure," http://www.counterpane.com/pki-risks.html

[25] R. Rivest, "The MD5 Message-Digest Algorithm," http://www.ietf.org/rfc/rfc1321.txt

[26] See http://www.cs.umd.edu/~waa/1x.pdf

[27] See work in progress—P. Funk and S. Blake-Wilson, "EAP Tunneled TLS Authentication Protocol (EAP-TTLS)," http://ietf.org/internet-drafts/draft-ietf-pppext-eap-ttls-00.txt

[28] W. Simpson, "PPP Challenge Handshake Authentication Protocol (CHAP)," http://www.ietf.org/rfc/rfc1994.txt

[29] See work in progress—H. Haverinen, "GSM SIM Authentication and Key Generation for Mobile IP," http://ietf.org/internet-drafts/draft-haverinen-mobileip-gsmsim-02.txt

[30] See work in progress—J. Carlson, et al., "EAP SRP-SHA1 Authentication Protocol," http://ietf.org/internet-drafts/draft-ietf-pppext-eap-srp-03.txt

[31] See work in progress—L. Salgarelli, "EAP-Shared Key Exchange (EAP-SKE); A Scheme for Authentication and Dynamic Key Exchange in 802.1x Networks," http://ietf.org/internet-drafts/draft-salgarelli-pppext-EAP-SKE-00.txt

[32] See work in progress—http://www.ietf.org/internet-drafts/draft-josefsson-pppext-eap-tls-eap-02.txt

[33] See http://www.cisco.com

# Chapter 9

# Building Secure Wireless Local Area Networks

*Colubris Networks*

Ubiquitous network access without wires. This is the powerful drawing card for the deployment of wireless networking technology. But, for many, this powerful advantage is seen as double-edged—providing increased flexibility and ease of use on one side, tempered by heightened security risks on the other. Essentially, the inability to physically secure a wireless network is considered to be its Achilles' heel. But this conclusion is flawed because it places the burden for security at the physical layer instead of at the network layer, where it makes much more sense. To dispel the wireless security myth, this chapter explains how to protect data with a secure wireless implementation.

## 9.1 Understanding Wireless Security Challenges

In addition to the great benefits wireless networks provide, enterprises that deploy this technology must understand and deal with the security issues related to radio transmission. Radio waves cannot be controlled, and they travel freely through most physical barriers, easily spreading confidential data beyond the walls of an office or home. With the transmission power of today's devices, ranges of 300 ft or more are common. If not handled properly, this potentially creates a major security hole in a network. However, this lack of security is not limited to wireless technologies. Data transmission technologies that rely on cables are not any more secure. They only happen to be easier to protect with physical barriers. Should the physical barriers be broken, then the actual security of the network can easily be compromised. For example, Ethernet LANs or dial-up Internet connections are both vulnerable to simple cable tapping. Because wireless technology has no inherent physical protection, it forces us to take a more critical look at current network security practices and to acknowledge their weaknesses. Once this is done, we can create solutions using existing proven technology.

## 9.2 Finding a Solution

The fact is that today most local area networks function without data security at the physical or logical link layers, as do the majority of dial-up and broadband Internet connections. They are generally only protected from intruders through some form of user authentication or Internet firewall. When data security is required, it is usually implemented at layer 3 or above. For example, transactions on the Web are secured with HTTPS (SSL). Remote access to corporate networks is secured with Virtual Private Networking (VPN). One of the main reasons for this is that implementing security at the physical layer is not always practical because a logical connection between two devices may be carried across more than one physical link. Providing end-to-end security between the two end points of a connection is more desirable because it functions independently of the underlying data transport. This type of security is best implemented at layer 3, the IP layer, because an IP datagram is the smallest addressable unit of data that is carried on an IP network. If we view wireless technology in this context, the approach shifts from trying to encrypt the radio transmissions (as is done by most wireless access points using Wired Equivalent Protection, or WEP) to creating secure end-to-end connections between stations. Currently, the most flexible method for doing this is to use an access point that integrates VPN.

## 9.3 The VPN Solution

VPN technology provides the means to transmit data securely between two network devices over an unsecure data transport medium. It is commonly used to link remote computers or networks to a corporate server via the Internet. However, it is also the ideal solution for protecting data on a wireless network. VPN works by creating a *tunnel* on top of a protocol such as IP (Fig. 9.1). Traffic inside the tunnel is encrypted and totally isolated.

VPN technology provides three levels of security, user authentication, encryption, and data authentication:

- Authentication ensures that only authorized users (over a specific device) are able to connect, send, and receive data over the wireless network.
- Encryption offers additional protection because it ensures that even if transmissions are intercepted, they cannot be decoded without significant time and effort.
- Data authentication ensures the integrity of data on the wireless network, guaranteeing that all traffic is from authenticated devices only.

By implementing VPN technology wireless networks become more secure than their unprotected wired counterparts, and can be used to solve even the most mission-critical networking challenges without security concerns.

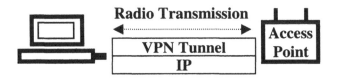

**Figure 9.1.** VPN technology can be used to create a secure tunnel over an unsecure protocol like IP.

## 9.4 Implementing VPN Security

Applying VPN technology to secure a wireless network requires a different approach than when it is used on wired networks. This is due to three factors:

- The inherent repeater function of wireless access points automatically forwards traffic between wireless LAN stations that communicate together and that appear on the same wireless network.
- The range of a wireless network will likely extend beyond the physical boundaries of an office or home, giving outsiders the means to compromise the network.
- The ease with which wireless networking solutions can be deployed, and their scalability, makes them ideal solutions for many different environments. As a result, implementation of VPN security will vary based on the needs of each type of environment.

### 9.4.1 Enterprise

In business environments, total security of the wireless network is crucial. This can be impossible to achieve with wireless solutions that rely exclusively on an external server for all VPN functionality. A security hole is created because access must be granted to the wireless network to enable computer users to reach the VPN server and log in (Fig. 9.2). Traffic flow on the wireless network cannot be completely secured.

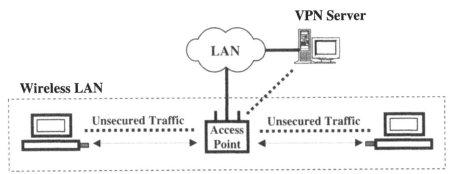

**Figure 9.2.** Wireless access point with an external VPN server.

To make effective use of VPN technology, the access point must have its own VPN server, or at the very least be VPN-aware. A VPN-aware access point only accepts and forwards VPN traffic to an external VPN server, discarding all other traffic (Fig. 9.3). Both implementations provide complete security for the network, because the access point will not allow wireless traffic *outside* of a VPN unless that traffic is to establish a VPN.

### 9.4.2 Public Access

In public access applications (airports, hotels, convention centers), the wireless service does not need to provide more security than what is offered by traditional dial-up services. This means that secure authentication for accounting purposes (usually via a centralized RADIUS server) achieved over an upstream VPN connection needs to be complemented by a wireless service that provides the ability to transparently support user-provided security on demand (typically a VPN connection to the corporate VPN server). Protection of individual stations from one another requires that the repeater function be disabled (Fig. 9.4).

### 9.4.3 Home/SOHO

Home/SOHO users may only need *moderate* protection for local traffic on the wireless network. This can usually be satisfied by implementing WEP encryption. However, for communication with a remote corporate network, it is important that the access point supports VPN security in "pass-through" mode or by embedding a

VPN client that can be shared between some of the devices connected to the wireless LAN (Fig. 9.5).

*Note*: This method does not address security concerns regarding data exchanged between client devices attached to the same wireless LAN.

### 9.4.4 Enhancing Wireless Security

The benefits of VPN technology can be enhanced by combining it with other security features. For example:

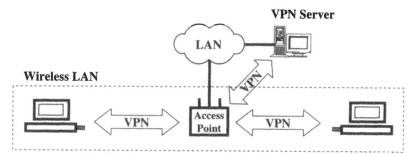

**Figure 9.3.** Wireless access point with integrated VPN server.

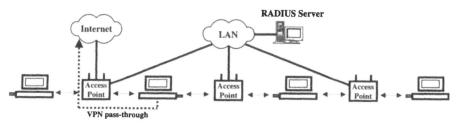

**Figure 9.4.** Multiple access points covering a large public area.

- Token-based authentication: Once a network is secured with VPN, user authentication can be strengthened. Additional verification of user identity can be implemented through hardware-based password generation (Entrust, RSA/SecurID, VASCO, etc.),
- VPN profiles: RADIUS or LDAP services can be used to manage individual user profiles. White/black lists can be created to control access to specific network resources or subnets.

## 9.5 Conclusion

The issue is not *Are wireless networks less secure?* but rather *What are the best methods to secure a wireless network?* Adopting the use of Virtual Private Networking leads to the application of a sound strategy based on strong user authentication and encryption achieved at the network (IP) layer without limiting the benefits of wireless LANs.

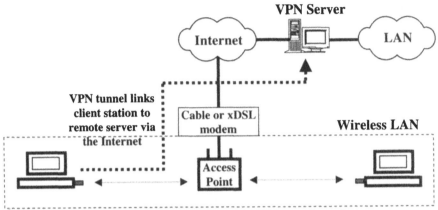

**Figure 9.5.** Access point with VPN pass-through.

# Chapter 10

# Wireless LAN for Mobile Operators

Philippe Laine
*Alcatel*

Wireless LANs will not replace mobile networks, but they provide an interesting complement, allowing broadband Internet access in selected hot spots. Mobile operators can offer their corporate customers a bundled offer of GPRS plus wireless LAN services. They will manage service subscription as well as customer care and billing. This could happen *now* with a simple, cost-effective solution based on existing equipment.

## 10.1 Internet over Mobile Networks

### 10.1.1 Global System Mobile

Since its inception, Global System Mobile or GSM (one of the second-generation mobile cellular systems) has been offering a connectionless packet service. This service, called Short Message Service (SMS), has been very successful in recent years, although it is limited to messages containing fewer than 160 characters. Recently, another solution to access data from a mobile phone (E-mail, Web browsing, file transfer, etc.) was proposed. This solution is Circuit Switched Data (CSD), which is similar to what is done in the fixed wireline network, where a dial-up modem is used to access the Internet. CSD offers a very limited throughput, only 9.6 kbit/s. The limitation of CSD led to the standardization of the General Packet Radio Service (GPRS).

### 10.1.2 General Packet Radio Service

GPRS is the first technology for mobile Internet (Fig. 10.1). GPRS is designed around a number of guiding principles:

- *Always-on*: End-to-end packet switching allowing a user to send or to receive information at any time;
- *High bit rates*: An actual bandwidth roughly equivalent to a wireline modem, i.e., around 40 kbit/s;

- *Improved usage of radio resources:* Several radio channels can be allocated to a single user and/or the same radio channel can be shared between several users;
- *Simultaneous voice call and data transfer:* Data can be sent or received during a circuit call;
- *Billing based on volume:* Subscribers billed on number of bytes received or sent whatever the duration.

GPRS keeps the same radio modulation, frequency bands, frequency hopping techniques, and TDMA frame structure as the GSM standard. A new functional network entity, the Packet Control Unit (PCU), is required in the Base Station Subsystem (BSS) to manage packet segmentation, radio channel access, automatic retransmission, and power control. The major new element introduced by GPRS is an overlay Network Subsystem (NSS) that processes all the data traffic. It comprises two network elements:

- *Serving GPRS Support Node (SGSN):* which keeps track of the location of individual mobile stations and performs security functions and access control;
- *Gateway GPRS Support Node (GGSN):* which encapsulates packets received from external packet networks (IP) and routes them towards the SGSN.

Enhanced Data Rate for Global Evolution (EDGE) improves GPRS by introducing a new radio modulation scheme that triples the bandwidth offered by GPRS.

### 10.1.3 Universal Mobile Telecommunications System

The Universal Mobile Telecommunications System (UMTS) is one of the systems retained for the third-generation (3G) mobile systems. The UMTS network consists of two independent subsystems connected over a standard interface (Fig. 10.2). These elements are:

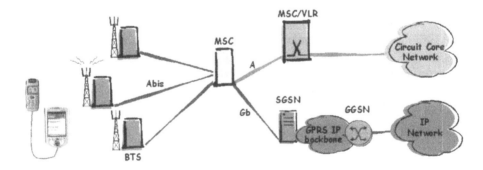

**Figure 10.1.** A GPRS network.

- *UMTS Terrestrial Radio Access Network (UTRAN)* composed of Node B and Radio Network Controller (RNC). Node B is equivalent to the GSM BTS, and RNC is equivalent to the GSM BSC,
- *UMTS Core Network*, which is similar to the GSM/GPRS NSS.

The radio is based on Wideband Code Division Multiple Access (WCDMA). Two different but related radio transmission modes are used:

- *Frequency Division Duplex (FDD)*, adapted for symmetric traffic (such as voice or video telephony);
- *Time Division Duplex (TDD)*, better suited to asymmetric traffic (such as Internet surfing).

With UMTS, users will get, in optimal condition, up to 400 kbit/s in FDD mode and up to 2 Mbit/s in TDD mode.

The first implementation of UMTS, which is based on Release 3 of the UMTS specification, reuses the entities already present on the GSM/GPRS NSS:

- *Packet Switch (PS)* is an evolution of the GPRS SGSN/GGSN;
- *Circuit Switch (CS)* is an evolution of the GSM switch.

### 10.1.4 Wireless LAN

Wireless LAN encompasses several technologies (Fig. 10.3):

- *IEEE 802.11b* leads the current deployment. 802.11b operates in the 2.4-GHz band and provides an actual throughput of about 5.5 Mbit/s. Since 2000, the Wireless Ethernet Compatibility Alliance has been certifying interoperability of 802.11b products from different vendors;
- *IEEE 802.11a*, an evolution of 802.11b, operates in the 5-GHz band and offers an actual throughput of about 30 Mbit/s;
- *ETSI HiperLAN2* also operates in the 5-GHz band, delivering an actual throughput of about 40 Mbit/s. HiperLAN2 solves many issues associated with 802.11a, in particular interference, security, and QoS.

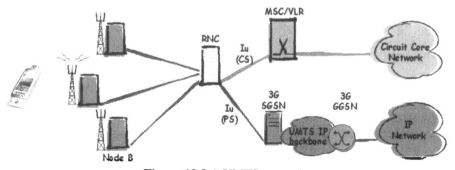

**Figure 10.2** A UMTS network.

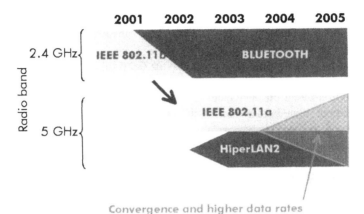

**Figure 10.3.** Wireless LAN evolution.

Nevertheless, the success of HiperLAN2 in the US is uncertain. The convergence of 802.11a and HiperLAN2 technologies is possible over the long term. Working groups at ETSI and IEEE started discussions recently. 802.11b products share the same frequency band (2.4 GHz) as HomeRF and Bluetooth products. If there are serious concerns over long-term prospects for HomeRF technology (a combination of simplified 802.11 and DECT targeting the residential segment), then Bluetooth will certainly be a great success in the years to come. The interference created by Bluetooth devices may reduce the throughput of 802.11b products to around 1–2 Mbit/s. When the number of Bluetooth devices becomes significant, wireless LAN products are likely to migrate to the 5-GHz band.

### 10.1.5 Hot Spots Coverage by Mobile Operators

The UMTS TDD mode is composed of two versions. One version, the UMTS High Chip Rate-TDD (HCR-TDD), is specifically dedicated for hot spots coverage. HCR-TDD cannot be used if the terminal is moving rapidly (in a car or train, for example). The other version, UMTS Low Chip Rate-TDD (LCR-TDD), or TD-SCDMA can be used in any type of environment. Both TDD versions offer a maximum throughput of 2 Mbit/s and, like FDD, offer extended roaming, security, and QoS support with several bearer classes. Whatever the technology that is selected, the TDD mode is available only in several years' time. In the meantime, to cover hot spots, operators may consider using an alternative technology such as public wireless LAN.

### 10.1.6 Road Map

GPRS is currently deployed in most European GSM networks and is progressively being commercialized. The EDGE upgrade will start in 2002, mostly in the US during the first phase. The installation of the first UMTS system is planned for the year 2002, but the commercial availability will start only during 2003. Currently, only 802.11b wireless LAN products are available. 802.11a products will become available in early 2002 and HiperLAN2 products slightly later.

## 10.2 Public Wireless LANs

### 10.2.1 Current Wireless LAN Use

Wireless LANs are designed for use inside corporate premises, but new applications are currently developed in residential (e.g., ADSL routers combined with wireless LANs as shown in Figure 10.4) and public environments. Some applications of wireless LANs in the public environment include:

- Local ISPs, mostly in the US, offering coverage of small cities with wireless LANs;
- Some cities (e.g., Seattle or Canberra) offering wireless LAN community services that are directly managed by the members of a community;
- Some companies (e.g., Wayport, MobileStar, etc.) proposing nationwide wireless LAN access in major airports and hotel chains.

This last service is the only one actually competing with mobile operators' offering, whereas others should be compared with Fixed Wireless Access. Current fee structures charged to the end users vary:

- Wayport charges a fee from $4.95 to $7.95 per day and per location;
- MobileStar offers subscription rates ranging from $15.95 for 200 minutes up to $59.95 for unlimited access;
- In Europe, Telia HomeRun offers similar services in Sweden and Norway but charges slightly higher fees.

### 10.2.2 Which Terminals for Public Wireless LAN Access?

Today, PCMCIA cards and Compact Flash cards are used to connect laptops and PDAs to wireless LANs (Fig. 10.5). These cards are on sale for $100/200. Some vendors are also proposing that PCs or PDAs include an embedded wireless LAN interface. However, even if they become widely available and cheap, these cards have a major problem—they are not optimized to reduce power consumption. For example, the operation of a connected PDA is currently restricted to about one hour. Although smartphones with embedded wireless LAN capabilities are also envisaged in the coming year, PDAs are the best-adapted devices for wireless LAN access in the public area. The PDA market already stands at almost 4 million units per year in Western Europe.

**Figure 10.4.** ADSL modem with embedded wireless LAN access point.

**Figure 10.5.** Wireless LAN PCMCIA card, PDA, and compact flash card.

## 10.3 Public Wireless LAN for Mobile Operators

There are some good reasons for deploying a wireless LAN:

- Wireless LAN technology will certainly be a key part of future networks because of the high bandwidth-to-cost ratio;
- Deploying wireless LAN technology enables mobile operators to offer broadband data services in hot spots, thereby complementing 3G/2.5G technologies;
- Deploying such a service helps to retain and acquire corporate customers, increasing the average revenue per month (ARPM) as the mass market saturates;
- Wireless ISPs (WISPs) have not strongly emerged in Europe as yet, but if this happens, WISP could jeopardize the mobile operator corporate customer market;
- The partnership with owners of hot spots could be simplified, thanks to the experience in site acquisition for GSM or UMTS networks.

Conversely, there are some other good reasons not to deploy a wireless LAN:

- Business models for public wireless LANs are not proven. Today's first business results of public wireless LAN are not convincing. Although not based on wireless LAN, the Metricom story is interesting. Metricom offers the Ricochet service in 17 major US cities (including New York, Los Angeles, San Francisco, and Seattle). Ricochet provided an unlimited wireless Internet access at a rate up to 128 kbit/s for about $75/month. However, Ricochet attracted only 40,900 subscribers, and the company filed for Chapter 11 bankruptcy in August 2001. Some reasons for this failure are proprietary technology, discontinuous coverage, and lack of service demands;

- Revenues do not appear to be sizable because the addressable end user market is currently limited to laptop users. Market adoption of PDAs with built-in wireless LAN interface is unclear today. In addition, a couple of years could be required to solve the power consumption problem;
- The unlicensed 2.4-GHz band will become rapidly crowded. 5-GHz products seem more appropriate to offer a good quality of service, but these products are not currently available;
- Partnership with a lot of owners of hot spots is required to achieve a critical mass of access zones.

Although the major threat is the lack of service demands, this is also a risk for UMTS.

### 10.3.1 Should Mobile Operators Invest in Wireless LAN?

For mobile operators, the move toward being a wireless LAN operator is not a proven business opportunity. However, to block the entrance of new players and to prepare for the availability of 5-GHz products, it makes sense to start a limited activity in selected locations and test the market response. ABN AMRO state in their last study about wireless LANs from August 2001 their belief that "public wireless LAN's best prospect for growth comes from a symbiotic relationship with 2.5G/3G mobile telephone networks."

### 10.3.2 Which Business Model?

Hot spots usually belong to companies such as airports or hotel chains. It makes sense for these companies to deploy a wireless LAN, to use it for their own applications, and to sell the unused capacity to wireless LAN operators. These companies can manage the installation and operation directly. Alternatively, they can request a wholesale access provider to do it for them. In this model, mobile operators are not deploying the access network but are buying capacities on networks deployed by others. Mobile operators can eventually establish a partnership with an ISP or a mobile portal to market the service. Synergy with current data service is essential for mobile operators. This can be done through a bundled offer comprising GSM/GPRS phone + PDA with wireless LAN or wireless LAN card. The mobile operator will manage customer care and billing. The end users will pay the usage of the wireless LAN to the mobile operator (Fig. 10.6).

## 10.4 Coupling of Wireless LAN and Mobile Networks

Several couplings between the wireless LAN network and the mobile network are possible.

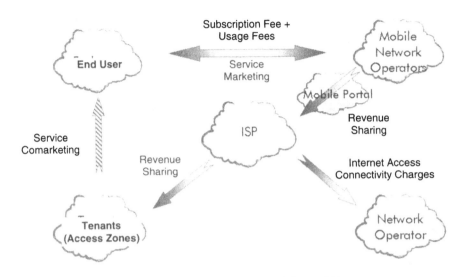

**Figure 10.6.** Wireless LAN business model for mobile operator.

### 10.4.1 Tight Coupling

Tight coupling is the only solution that brings a seamless handover between a wireless LAN and the mobile network, offering the same level of security as UMTS (Fig. 10.7). It requires the standardization of a simplified Iu interface (Iu is the interface between the UMTS Radio Access Network and the Core Network). The ETSI project called Broadband Radio Access Network (BRAN) is working on this specification for HiperLAN2. Tight coupling requires specific access network equipment and a wireless LAN terminal with an embedded Security Identity Module (SIM) card.

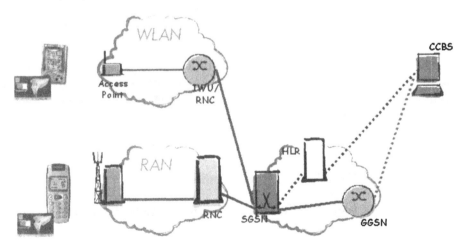

**Figure 10.7.** Tight coupling.

## 10.4.2 Loose Coupling

Loose coupling offers the same security benefits as tight coupling while requiring less standardization effort (Fig. 10.8). The link between a wireless LAN and the mobile network is performed between the Authentication, Authorization, and Accounting (AAA) server and the Home Location Register (HLR). The HLR stores the current location of mobile subscribers and the list of services to which they have access rights. The ETSI BRAN committee is also working on the specification of this interface for HiperLAN2. Unlike tight coupling, loose coupling does not require specific access network equipment. The wireless LAN terminal could include a SIM card.

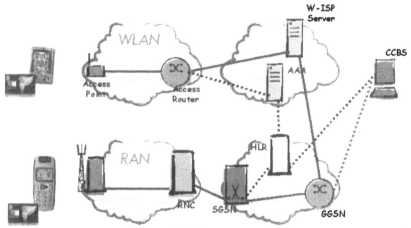

**Figure 10.8.** Loose coupling.

## 10.4.3 Open Coupling

Open coupling is a simple solution that does not require standardization (Fig. 10.9). The link between the wireless LAN and the mobile network is performed at Customer Care and Billing System (CC&BS) levels. The AAA server sends information related to the usage of the wireless LAN network to the mobile network CC&BS. This solution does not employ the mobile network security mechanisms. All access equipment and wireless LAN interface cards are standard commercial products.

## 10.4.4 Proposed Solution

Tight and loose couplings offer security but require specific equipment, meaning higher investment for operators and end users. New interfaces also need to be standardized. Open coupling requires no specific equipment, needs a limited investment, and can be deployed today. To test the market, open coupling seems more adapted because of its readiness and its cost. In the long term, the evolution of the network toward loose coupling is linked to the availability of terminals with wireless LAN capabilities and embedded SIM card (Fig. 10.10).

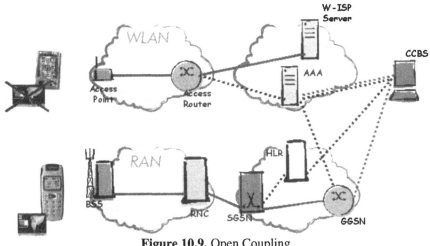

**Figure 10.9.** Open Coupling.

## 10.5 Issues

Several issues should be studied carefully before introducing wireless LAN in a mobile network.

### 10.5.1 European Regulation

In France, Italy, and the United Kingdom, public networks are not permitted. In some other countries, a license is required for public usage. Among these countries are Belgium (€12500) and Germany (m€ 5). In most countries, the output power is limited according to ERC recommendations:

- 2.4-GHz band: Up to 100 mW;
- 5-GHz band: Up to 200 mW in the bands reserved for HiperLAN2; up to 25 mW in bands opened to other technologies.

It should be pointed out that most of the 5-GHz band is reserved for HiperLAN2 (455 MHz) or any system using Dynamic Frequency Selection (DFS). Only 150 MHz is available for other technologies, including 802.11a.

### 10.5.2 Billing

When introducing GPRS, in most cases, operators do not modify their existing billing system. The preferred solution is a separate rating engine that collects GPRS Call Data Records (CDR) from the Charging Gateway. The rating engine then sends the rated file to the existing billing system. The GPRS rating engine can be upgraded to collect wireless LAN CDR from the AAA server. Instead of a charging gateway, it should be noted that the AAA server is only issuing information based on time usage.

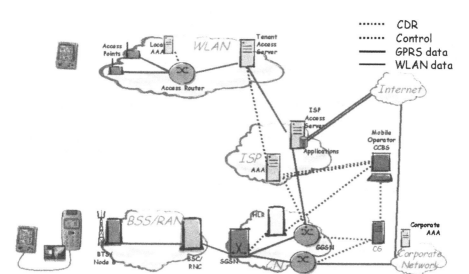

**Figure 10.10.** Proposed solution.

### 10.5.3 Security

Security on a wireless LAN is normally built with the Wired Equivalent Privacy (WEP) protocol, which provides confidentiality and integrity of the wireless traffic. However, as it currently stands, WEP has been shown to contain flaws and does not meet the claimed security level. Moreover, the lack of a key management protocol has also brought about major concerns (e.g., a secret key is chosen manually and, if poorly chosen, can easily be cracked using a list of common passwords). Finally, the use of the Dynamic Host Configuration Protocol (DHCP) needed for mobility services does not provide strong security services.

In the longer term, a new mechanism will have to be made available to have a very secure protocol at the wireless network level. Solutions are currently under development to have the end user authenticated when connecting to the wireless LAN using DHCP and to improve DHCP security. In the short term, it is necessary to build architectures using existing technologies. An immediate solution to build a secure communication is to make use of an IPSec-based Virtual Private Network (VPN), at least when setting up wireless LAN access to corporate networks. Having received IP connectivity, the end user is authenticated when setting up the IPSec tunnel to the (visited or home) service provider or corporate network. Terminals must embed an IPSec client to be able to set up the VPN.

### 10.5.4 Roaming

Roamers (users visiting a wireless LAN not managed by their own operator) should be able to access the Internet through this network and be billed through their operator. This is typically achieved using some AAA mechanism (e.g., RADIUS servers). For a roamer, the visited AAA server relays authentication information (using RADIUS) to the home AAA server. The home AAA server can either perform the user's authentication or relay the RADIUS packet to the corporate AAA server. If authentication is successful, the visited access server assigns an IP address to the wireless user (i.e., the wireless user is part of the visited network IP realm). Interactions between the AAA servers can be protected using IPSec or Secure Sockets Layer (SSL). Once this is done, as in the standard case, the user should set up an IPSec tunnel with his corporate network.

Two possibilities are offered to mobile operators to interconnect their wireless LANs:

- Peer agreements: Two operators establish a link between their AAA servers. These bilateral agreements already exist for GSM and GPRS roaming.
- Internet roaming: Some companies are today providing global Internet roaming, allowing seamless interconnections and financial settlements.

### 10.5.5 Applications

The service will be successful only if mobile operators include a complete environment for messaging and business:

- Internet/Intranet access;
- E-mail, unified messaging, assistant organizer, agenda, directory;
- Access to specific information (e.g., ticketing and reservation information).

These applications should be available seamlessly from mobile and wireless LAN networks. Applications requesting large bandwidths could possibly be offered only over wireless LAN network (e.g., download of music files, high-quality video streaming, or on-line interactive games). In addition, services can be developed over mobile networks specifically for wireless LAN users:

- Indication of wireless LAN service availability in the area;
- To increase security, wireless LAN service is made available only if the user GSM handset is also connected in the area;
- One time password sent through SMS.

## 10.6 Conclusions

Wireless LANs will not replace mobile networks, but they provide an interesting complement, allowing broadband Internet access in selected hot spots. Mobile operators can offer their corporate customers a bundled offer of GPRS plus wireless LAN services. They will manage service subscription as well as customer care and billing. This could happen *now* with a simple, cost-effective solution based on existing equipment.

# Chapter 11

# Wireless LAN Architecture for Mobile Operators

Juha Ala-Laurila, Henry Haverinen, Jouni Mikkonen, Jyri Rinnemaa
*Nokia Mobile Phones*

The evolution of Internet- or IP-based office applications has created a strong demand for public broadband wireless access offering capacity beyond current cellular systems. A wireless LAN is able to support broadband data rates in indoor environments but does not include support for operator-driven public access. In addition, most commercial public wireless LAN solutions only provide modest authentication and roaming capability compared with traditional cellular networks. This chapter describes a new wireless LAN system architecture that combines the wireless LAN radio access technology with the mobile operators' SIM-based subscriber management functions and roaming infrastructure. In the defined system, wireless LAN access is authenticated and charged with the GSM SIM, which is a widely deployed mobility platform. This solution supports roaming between cellular and wireless LAN access networks and is one step toward an all-IP network architecture. The first release of the system has been implemented and verified in an actual mobile operator network.

## 11.1 Introduction

Seamless access to modern office tools is one of the most valuable assets for mobile business professionals today. Most corporate information systems and databases can be accessed remotely through the Internet backbone, but the high bandwidth demand of typical office applications (e.g., large E-mail attachment downloads) often exceeds the transmission capacities of cellular networks. Mobile professionals are looking for a public wireless access solution that can cover the requirements of data-intensive applications and enable smooth on-line access to corporate data services.

Wireless local area networks that employ radio technology differ from cellular technology by having a superior bandwidth compared with cellular technology and by permitting a license-exempt use. For example, the Wi-Fi[TM] wireless LAN standard (IEEE 802.11b) offers a maximum peak data rate of 11 Mbit/s (typical application layer peak data rate is 6.5 Mbit/s) [1, 2]. This is in contrast to a General Packet Radio Service (GPRS) handset that offers a data rate of up to 172 kbit/s (typical 42 kbit/s) and the third-generation mobile terminal that provides up to 2 Mbit/s (typical 144 kbit/s). From the end users' perspective, the performance difference is significant. Furthermore, 802.11b operates over the unlicensed 2.4-

GHz frequency band, which allows new types of wireless access networks to be created. For instance, 802.11b wireless LANs were initially deployed as private wireless networks in homes or corporation premises. In the second wave, the same technology is extended into public access environments that benefit visiting guests. Target places for public access wireless LAN services include airports, railway stations, hotels, business parks, and office buildings, where most mobile laptop users typically work (Fig. 11.1).

Most wireless LAN terminals are laptops or PDAs with separate wireless LAN network adapters. Market analysts forecast major growth for the wireless LAN PC card market in 2001. This is partly a result of leading laptop manufacturers starting to integrate wireless LAN devices in their high-end models during 2001. Even though the terminal business is driven primarily by the PC segment, it is expected to expand further with wireless LAN Web pad, phone, and PDA devices.

Business professionals' natural requirement for broadband data access and the rapid growth in wireless LAN terminal penetration create a business opportunity for mobile operators to extend their services to cover wireless LAN access. Today, there are about 50 million mobile laptop users, of which roughly 30 million have GSM subscriptions [3]. Wireless LANs can complement a mobile operator's traditional wide-area GPRS and GSM services by offering a wireless broadband solution for indoor environments that may be extended by the user in a private deployment at home. The key question is how the mobile operators can exploit their strengths (e.g., large customer base, cellular infrastructure investments, and well-established roaming agreements) when operating wireless LANs. The authors have been involved in the development of a new wireless LAN system architecture that provides one solution for mobile operators. The defined solution is called the operator wireless LAN system. The system enables IP service roaming between wireless LAN hot spots by using cellular subscriber and roaming management. This integration approach is often called a "loose interworking" model.

The operator wireless LAN supports any wireless LAN device with a GSM Subscriber Identity Module (SIM) card reader and includes the operator wireless LAN client software. A prototype of the solution was successfully piloted in an actual mobile operator network in 2000. The observations and lessons learned from the trial proved the usability of the concept. This resulted in the first commercial system being launched in July 2001. The first commercial release of the operator wireless LAN solution did not require any special support from the wireless LAN access layer. It was based entirely on standard IP-based protocols. This approach is designed to support backward compatibility during the evolutional releases of the solution, which will make use of the new IEEE 802.11i standard [5]. The IEEE standard defines new wireless LAN level authentication and access control mechanisms (IEEE 802.1x [4]) and enhances the wireless LAN security by adding improved encryption schemes. One of the benefits enabled by 802.11i, when it becomes available, is dynamic air interface packet security with encryption keys provided by the network.

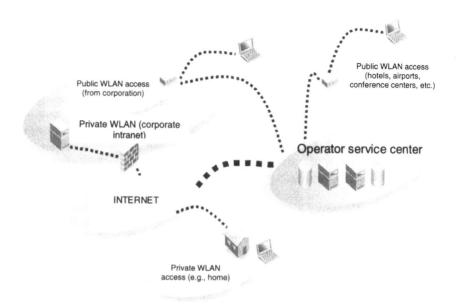

**Figure 11.1.** Wireless LAN in the private and public environments [19].

Most of the wireless LAN development is targeted for private operations (e.g., in homes or corporations). This chapter focuses on describing how wireless LAN access can be seamlessly integrated with cellular networks. The system architecture, its main components, and their functions are described in detail. The focus is on the GSM SIM-based authentication, roaming, and billing mechanisms as well as the first-release system architecture based on IP-level authentication. The chapter concludes with an overview of future evolution prospects and the status of standardization.

## 11.2 Operator Wireless LAN Overview

### 11.2.1 Design Objectives

The future mobile operator network is expected to be a combination of several radio communication technologies such as GSM/GPRS and wireless LAN. A single subscriber identity should be used in all access networks to enable smooth roaming and seamless service availability. The operator wireless LAN system should maintain compatibility with the existing GSM/GPRS core network roaming and billing functions [6, 7], which minimizes the number of modifications needed in the GSM core equipment and the required standardization effort. A SIM is a natural choice for wireless LAN subscriber management because it is widely deployed and enables roaming in existing GSM/GPRS networks. Mobile business professionals often own a GSM subscription, and it is convenient to receive only one bill from a trusted party (i.e., the cellular operator) providing all mobile services (i.e., voice and data).

In the first phase, the focus of wireless LAN business is on data centric applications. Thus the operator wireless LAN system should be optimized for terminal initiated IP data services, which also keeps the system complexity manageable. The evolution to all-IP services in the wireless LAN domain should be straightforward when the mobile operator all-IP service infrastructure becomes available. This all-IP network will then add network-initiated services and real-time services for the supported access networks. To minimize installation costs and complexity, the operator wireless LAN should utilize the existing GPRS charging system. GPRS operators have defined a mechanism for exchanging billing data between operators' networks [8]. This same mechanism can be used for transmitting wireless LAN charging records.

## 11.2.2 Operator Wireless LAN System Architecture

The operator wireless LAN system architecture consists of the public LAN access network and the cellular operator site that communicates over the IP backbone. The main design challenge is to transport standard GSM subscriber authentication signaling [6] from the terminal to the cellular site using the IP protocol framework. This is the first time that the GSM protocols are integrated commercially with standard IP protocols. The operator wireless LAN system comprises four key physical entities (Fig. 11.2):

- Authentication Server;
- Access Controller;
- Access Point;
- Mobile Terminal.

The system architecture resembles the GPRS network. In fact, each system component has a counterpart in the GPRS network as illustrated in Table 11.1. The main difference from the GPRS architecture is that in the operator wireless LAN system, only the control signaling information is transported to the cellular core. User data packets are directly routed by the access controller (AC) to the IP backbone, which is used for accessing public and private services. Clearly, this architecture enables efficient routing of user data compared with GPRS, because the user's IP traffic does not have to be carried via the cellular core to the Internet backbone or user's corporate network. Consequently, the operator wireless LAN approach decreases the load of the cellular core.

The first release of the operator wireless LAN solution performs subscriber authentication at the IP level. The IP level authentication works as follows. First, a wireless LAN terminal associates with a wireless LAN access point, gets an IP address from the AC, and initiates network authentication by sending a dedicated authentication request to the AC. The AC relays the authentication request to the authentication server (AS), which implements a gateway between the access network and the GSM signaling network. The AS queries the GSM Home Location Register (HLR) for authentication data and performs user authentication using this information.

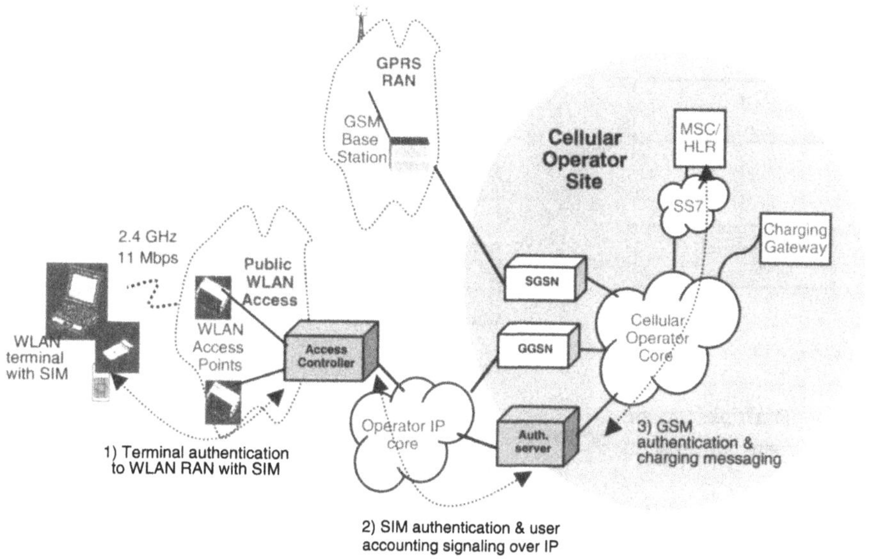

**Figure 11.2.** Overview of an operator wireless LAN system architecture.

## 11.3 System Elements

Figure 11.3 depicts, in more detail, the major elements and interfaces of the operator wireless LAN system.

### 11.3.1 Authentication Server

The AS is the main control point of operator wireless LAN subscriber management. A single entity may support several ACs and provide authentication/billing services for thousands of roaming users in different access zones. The AS communicates with the AC with the RADIUS authentication protocol [11], which is a de facto Authentication, Authorization, and Accounting (AAA) protocol employed by the IP industry. When the user disconnects, the AS receives the accounting data from the AC, converts it into a GPRS billing format [8], and issues the ticket to the cellular billing system.

**Table 11.1. Operator wireless LAN versus GPRS network elements.**

| Operator Wireless LAN | GPRS | Function |
|---|---|---|
| Authentication Server (AS) | SGSN | User authentication, access billing |
| Access Controller (AC) | GGSN | End point of IP packet network, IP address allocation |
| Access Point (AP) | BTS | Radio coverage |
| Mobile Terminal (MT) | Mobile Phone | End user device |

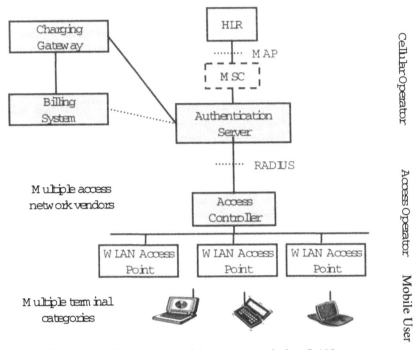

**Figure 11.3** Components of the operator wireless LAN system.

The AS hides the cellular infrastructure from the access network. It provides a gateway to the cellular core network elements, namely, the GSM Home Location Register (HLR) and the GPRS charging gateway. The AS sends standard GSM authentication signaling to the HLR with the SS7 signaling network that connects various operator networks together. The cellular network identifies the user with the GSM International Mobile Subscriber Identity (IMSI) code stored in the SIM card.

In the first system release, the AS was implemented with Windows NT. The AS was directly connected to the Nokia Mobile Switching Centre (MSC). An IP-compliant vendor specific protocol carries the authentication requests from the AS to the MSC. Figure 11.3 shows the applied configuration in which the MSC offers a GSM-specific Mobile Application Part (MAP) protocol interface. MAP is used for communication between the MSC and the HLR. The MAP interface is deployed as a redundant interface to avoid a single point of failure. The signaling traffic from the wireless LAN is handled by the MSC so that wireless LAN-related requests do not block GSM signaling. In subsequent commercial releases, the AS includes a native MAP interface and hence does not require the operator to use a Nokia-supplied MSC.

A predefined bit pattern in the HLR subscriber service profile indicates the wireless LAN service subscription. The AS always checks whether the roaming user is subscribed to the wireless LAN service. In the future, the service profile can be extended to include new features (e.g., the quality of service or QoS class of a wireless LAN subscriber). The current GSM standard does not define a wireless

LAN service. As a result, the rules for interpreting the bit pattern of the service profile are operator dependent.

### 11.3.2 Access Controller

The AC provides an Internet gateway between the Radio Access Network (RAN) and the fixed IP core. It allocates IP addresses to the mobile terminals and maintains a list of authenticated IP addresses. The AC acts as a traffic filter, monitoring the address of each incoming or outgoing IP packet and discarding packets that are generated from unauthenticated terminals. The AC separates the mobile terminals using the IP address and the unique wireless LAN MAC address. The MAC address verification ensures that a duplicate IP address cannot be used by a hostile user. The AC also gathers accounting information for billing purposes. The AC has been implemented on an IP router platform, more precisely, the Nokia IPSO IP330 series router.

### 11.3.3 Wireless LAN Access Point

The AP offers a wireless Ethernet link between the mobile terminal and the fixed LAN. APs are connected to the same LAN using the AC. The ACs are Wi-Fi™ compliant and support data rates of 1, 2, 5.5, and 11 Mbit/s [1, 2]. The typical coverage range of a single Wi-Fi™ AP is 50–100 m indoors. The coverage can be extended with directional antennas and radio network planning tools. The AP offers a shared radio interface. Consequently, the number of active terminals accessing the AP affects the perceived user data rate. If there are no other terminals in the coverage area of an AP, a single user may use the full 11 Mbit/s radio link rate. This is a significant difference compared with GPRS radio access [7], which is based on dedicated connections.

### 11.3.4 Mobile Terminal

The operator wireless LAN service is available for any terminal with wireless LAN radio access capability, SIM reader, and operator wireless LAN software module. Today, many different mobile terminal (MT) solutions are available in the PC environment, which is seen as the first terminal platform for the operator wireless LAN. The end user may deploy either a wireless LAN card with an integrated SIM reader or a wireless LAN card with an external/integrated smart card reader. It is also notable that high-end laptop models are starting to have integrated wireless LAN and smart card reader capabilities. A GSM/GPRS PC card used on a laptop PC with an integrated wireless LAN will be able to provide complementary wireless LAN and cellular data services—coverage area can be traded off with bandwidth requirements. Next, it is envisaged that both the cellular and the wireless LAN technologies will be integrated into the PDA segment, making high-bandwidth (hot spot) usage more convenient for mobile users.

Wireless LANs are initially planned for private access. Public access will create pressure to improve network discovery. This is because any given location may be covered by a number of wireless LANs, with different networks supporting private and public access. The wireless LAN standard does not indicate the type of access support (i.e., public vs. private). The simplest way to find a public access network is to test the networks by trial and error or to keep a database containing the names of public access networks. For practical reasons, the roaming operator wireless LAN terminal will need to detect the correct network by using predefined network profiles, which contain identifiers indicating the roaming partners' wireless LAN. When entering a new location, the terminal compares the names of available wireless LANs with the roaming profile and associates with the correct wireless LAN. The operator may distribute the profiles with a SIM card or a WWW server. This is the recommended approach to get started without any dependence on standardization changes.

## 11.4 System Operation

The need to maintain compatibility with existing wireless LAN devices and the cellular core implicitly leads to an architecture in which the necessary SIM-specific signaling messages are transported with the IP protocol. The IP approach makes the operator wireless LAN system independent of the wireless LAN standard, thereby allowing the same concept to be deployed for future 5-GHz wireless LAN systems (e.g., IEEE 802.11a and HiperLAN2).

The functional division between the AC and AS keeps the complexity of the core network manageable. The computing-intensive IP packet filtering and routing functions reside in the access network side, where the processing load can be distributed among a number of ACs. This improves system scalability and robustness. Figure 11.4 illustrates the resulting control plane architecture.

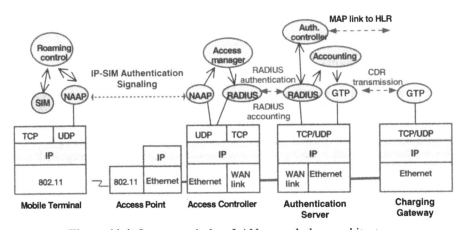

**Figure 11.4.** Operator wireless LAN control plane architecture.

The core component of the mobile terminal software is the Roaming Control module, which offers a graphical control user interface to roaming services. This module communicates with the SIM card. The operator wireless LAN-specific Network-Access Authentication and Accounting Protocol (NAAP) [9] encapsulates GSM authentication messages in IP packets. NAAP utilizes the Extensible Authentication Protocol (EAP) [10] over the connectionless User Datagram Protocol

(UDP) transport layer. The key component of the AC is the Access Manager, which controls IP routing and collects accounting statistics. The RADIUS protocol [11] carries SIM-specific authentication parameters within vendor-specific attributes while the accounting data are submitted with standard RADIUS accounting attributes [12].

The most important module in the AS is the Authentication Controller, which handles the RADIUS authentication messages and communicates with the GSM core. The Accounting module receives and stores the accounting information from the access network. The Accounting module interfaces with the GPRS charging gateway using the GTP' billing protocol [8]. The GTP' protocol has been derived from the GPRS Tunneling Protocol (GTP), which is used for packet data tunneling in a GPRS backbone network. No uniquely defined open GTP interface is available. Rather, there exist several versions of GTP as well as various billing data formats. The AS offers an open FTP interface, which can be used for fetching accounting data directly to various billing systems. The following sections describe the main system procedures, such as authentication, billing, and secure access to the corporate network.

### 11.4.1 Authentication

The core part of the operator wireless LAN system is the SIM-based authentication. The AC acts as a relay between the terminal and the AS. The authentication sequence is as follows: The first step is to activate the user terminal with the Personal Identity Number (PIN). The SIM card and user identity are secured with PIN code protection. When the authentication phase is activated, the software prompts the user for the PIN. This functionality is equivalent to GSM authentication and makes it impossible to use a stolen SIM card that has the PIN query enabled without knowledge of the correct PIN.

The phases of SIM-based authentication are illustrated in Figure 11.5. The steps are numbered both in the figure and the text below. Initially, the terminal locates the AC in the network by sending a NAAP solicitation message to which the AC replies (1). After receiving the IP address of the AC, the terminal sends the initial authentication request to the AC, where the IMSI is encapsulated into the Network Access Identifier (NAI) (2). NAI is a well-known part of the standard IETF AAA framework [13]. The domain part of the NAI identifies the home operator domains (e.g., IMSI@operator.com). The AC and RADIUS infrastructure use the domain part for relaying the authentication request to the correct AS (3). The AS requests GSM authentication information from the HLR with MSC connection (4). The GSM authentication information contains the RAND (random number), which the AS sends to the mobile terminal along with a message authentication code, which is calculated over the RAND (5, 6). The message authentication code authenticates the cellular network to the terminal, thereby enabling mutual authentication. The terminal calculates the message authentication code and compares it to the one received from the network (7). If the calculated message authentication code does not match with the received code, the terminal can suspect a fraudulent service and stop answering the authentication request. This mechanism makes it impossible for an attacker to gather RAND and Signed Response (SRES) pairs from the IMSI and thus makes it impossible to find out the secret key stored on the SIM card. The message authentication procedure utilizes the HMAC mechanism presented in Reference 14. Next, the terminal calculates the SRES with algorithms stored on the SIM card and calculates a message authentication code over the SRES (8). The terminal sends the response to the AC, which relays it to the AS (9, 10). The AS verifies the response by calculating the message authentication code over the SRES (11). The AS sends the resulting authentication code to the AC (12). If the authentication was successful, the AC sends the AS an indication that a new accounting session has been started (13). Finally, the AC enables routing for the terminal data packets and sends the acknowledgment to the terminal (14, 15).

The authentication procedure may be aborted for the following reasons:

- The IMSI is not known to the HLR;
- The wireless LAN service has not been subscribed;
- The access operator does not support roaming of the specific IMSI;
- The AS did not receive the security triplets (i.e., challenge/response pairs comprising RAND/SRES and secret key) from the HLR.

In these cases, the terminal remains unauthenticated, and the AC does not route the IP traffic of the terminal.

During the authentication procedure, the AS sends the terminal and the AC a session lifetime value, which indicates how long the authenticated session is valid. When the session lifetime expires in the AC or the terminal disconnects, the AC closes the terminal's connection. It also notifies the AS, which closes the accounting session for the terminal. The terminal must initiate a reauthentication sequence before the session lifetime expires if it wishes to continue the session. The reauthentication procedure allows the operator to grant an access, which is valid only for a certain period. This is a useful feature for prepaid systems as well as for an environment where charging can be bound to certain access time. This also allows the operator to authenticate the terminal periodically during an active session, just as in the GSM system.

### 11.4.2 Accounting and Billing

The AC monitors the data traffic and periodically sends information about traffic statistics to the AS. The following accounting methods are supported:

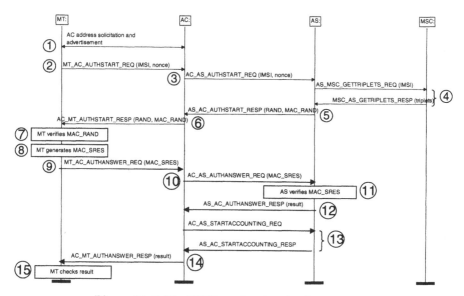

**Figure 11.5.** The SIM-based authentication sequence.

- Time-based: Connection start time and end time are recorded;
- Volume-based: The amount of transferred data is recorded;
- Flat rate.

The AS converts the accounting data to standard GPRS Call Detail Record (CDR) format. The wireless LAN CDRs are marked with a wireless LAN specific identifier code. More details on the CDR format can be found in Reference 8.

The AS verifies the received accounting data related to an authenticated terminal (standard IMSI is used for identification). This ensures that charging records are not generated for a connection that has not been properly authenticated. Furthermore, accounting data are protected with the shared secret mechanism in the RADIUS protocol [11]. This mechanism prevents fraudulent insertion of accounting data in the IP network because the AC and the AS that share the secret key will notice any alteration of data during transmission. The shared secret key is entered in the AC and AS during system configuration. The transferred data can also be encrypted using the IP Security (IPSec) protocol [15] between the AC and AS. Finally, the AS delivers the generated CDRs to the charging gateway or billing system. The explicit rules, which the connection is eventually billed, are configured in the back-end process of the operator billing system. Table 11.2 depicts the pricing models for two early public wireless LAN access offerings. Both models assume that the subscriber already has a terminal with wireless LAN radio access capability. Two kinds of charging schemes are offered to the customer: flat rate and time based. In the Telia HomeRun service's flat rate option, the charges for connection usage are included in the monthly fee. In the base pricing model, subscribers pay an entrance fee for the service, a monthly subscription fee, and per-minute charges for connection usage. The 24H option is similar to the flat rate, and includes charges for a 24-h period usage of the service. This option can be paid for before using the service; the subscription is kept activated for 24 h from the first log on. Sonera's service pricing for wireless LAN access is time based and is offered as part of the cellular data related service package.

### 11.4.3 Roaming to Foreign Wireless LAN Networks

Unlike most Internet service providers, mobile operators have the infrastructure to support roaming between different access networks as well as between operator networks. The mobile operators have agreed on how to exchange the authentication and accounting data between the roaming and the home operator. The existing operator billing systems are also capable of exchanging accounting information with each other. Figure 11.6 illustrates how the operator-to-operator roaming scenario works in the operator wireless LAN system.

First, the roaming mobile terminal associates with the foreign operator wireless LAN and initiates authentication by sending an authentication request to the AC, which relays it to the AS (1). The AS analyzes the IMSI and verifies that the operators have a valid roaming agreement for wireless LAN services. Next, the AS sends the authentication query to the correct HLR with the GSM SS7 network (2) [6]. The corresponding HLR responds with user profile and authentication triplets, and the authentication procedure is completed in the normal way. After the terminal disconnects, the AS sends the charging record to the foreign operator's billing system (3). The IMSI code indicates that the CDR is generated for a roaming terminal. The operator billing systems regularly communicate with each other and exchange GSM/GPRS and wireless LAN specific billing records generated by roaming users. Using this mechanism, the foreign operator's billing system relays the wireless LAN CDR to the user's home operator billing system, which finally submits the end user bill (4).

Because the operator wireless LAN solution makes use of the RADIUS protocol, all the roaming features of RADIUS are also available. Although the scale of the global GSM/GPRS roaming network surpasses current RADIUS networks, it may still be useful to supplement cellular roaming with RADIUS roaming. For example, a mobile operator and an Internet service provider can share their wireless LAN access networks and use RADIUS roaming to route the AAA packets between the access networks and the ASs. The ability to use both roaming methods, even simultaneously, makes the operator wireless LAN system the most flexible wireless LAN roaming solution in the marketplace.

**Table 11.2. Early pricing models for public wireless LAN access.**

| Fee name | Fee type | Price (USD) | Operator |
|---|---|---|---|
| Flat rate | Entrance | 46.0 | Telia [17] |
| Flat rate | Subscription per month | 140.0 | Telia |
| Base | Entrance | 46.0 | Telia |
| Base | Subscription per month | 28.0 | Telia |
| Base | Charges per minute | 0.2 | Telia |
| 24H | Charges per 24 h | 9.0 | Telia |
| Company data | Entrance | 7.3 | Sonera [18] |
| Company data | Subscription per month | 3.6 | Sonera |
| Company data | Charges per minute | 0.4 | Sonera |

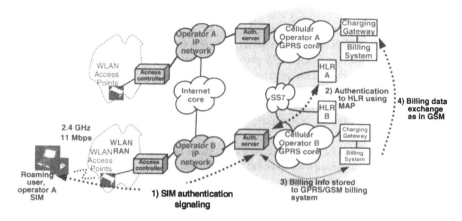

**Figure 11.6.** Wireless LAN roaming in a foreign operator network.

## 11.4.4 Secure Remote Access for Corporate Users

The target customers of the operator wireless LAN service are mobile business users who use wireless LAN extensions for accessing the corporate network. To guarantee the privacy of sensitive business information, an end-to-end encrypted connection must be established between the terminal and the corporate network. Typically, this is provided with a Virtual Private Network (VPN) server in the corporate network side, and corresponding VPN client software in the remote terminal [15]. The usability of remote access may further be improved by storing the VPN authentication certificate in the SIM. This makes it possible to protect the VPN keys and to launch seamless VPN authentication after SIM-based authentication when accessing the network. In this approach, the operator can offer both VPN and access service in close cooperation with the corporate information management department. It should be pointed out that a few VPN clients only support routable IP addresses instead of private IP addresses. This is a common constraint for all remote access services, because the operator has to allocate a large number of routable IP addresses for roaming users. The operator wireless LAN architecture requires the use of modern VPN products with private addressing support, which improves the scalability and usability of the public wireless LAN access system.

## 11.5 Scalability and System Robustness

Mobile operators impose extremely high error tolerance and resilience requirements on the networking infrastructure. To meet these requirements, the operation of the AP, the AC, and the AS must be reliable and the system must offer sufficient redundancy characteristics.

The critical part of the operator wireless LAN system is the fault tolerance of the AS because a single AS may serve several radio access networks and tens of thousands of mobile users simultaneously. A minimum installation of the operator wireless LAN consists of two ASs. Fault tolerance of the AS is supported with standard RADIUS proxy infrastructure (Fig. 11.7). The AC is connected to both proxies, one being the primary and the other the secondary RADIUS proxy. If the AS does not reply to the AC's messages, the AC sends the message to the secondary RADIUS proxy, which relays the messages to the secondary AS. Thus redundant IP routing infrastructure and RADIUS proxies ensure seamless switching between the ASs in case of an error. This approach allows the mobile operator to increase the system capacity by adding more ASs to the network. It must be noted that only interoperability-tested proxy servers should be deployed.

The risk of losing accounting data is minimized by updating the AS periodically. The operator may further improve system robustness by installing a secondary AC, which is used if the primary AC is broken or under heavy load. The AC platform also monitors its internal functionality. If the platform is not working correctly, it will send alarms to the network management system and/or automatically reboot. The redundancy of the AP is not extremely critical. If an AP is not functioning correctly, it will turn off its radio, after which the terminal automatically roams to a new AP.

## 11.6 Standardization Considerations and Future Prospects

Operator wireless LAN authentication protocols transport GSM SIM authentication information in EAP packets. EAP, specified in Reference 10, is a widely used authentication protocol originally designed to function in a Point-to-Point Protocol (PPP) authentication framework. We have specified a GSM SIM authentication scheme using EAP for operator wireless LAN. The scheme is being standardized in the Internet Engineering Task Force [16].

The IEEE 802.11 working group is currently standardizing enhanced wireless LAN security in Task Group I [5]. Part of the work concentrates on improving wireless LAN packet security with improved encryption and integrity protection. In addition, the task group specifies how IEEE standard 802.1x [4] can be used with wireless LANs. IEEE 802.1x specifies port-based access control on the various IEEE access media (e.g., 802.3 and 802.11) and uses EAP for authentication. Dynamic keys are distributed as part of the EAP authentication. The standardization work is still in progress, and the operator wireless LAN solution will support these new standards once they become available. Because EAP is used for subscriber authentication in IEEE 802.1x, the GSM SIM EAP type already used in the first release of operator wireless LAN will also be applicable to the IEEE 802.11i-based solution. The upcoming releases of the operator wireless LAN solution will support both IP-based authentication for legacy wireless LAN hardware and IEEE 802.11i authentication for new wireless LAN hardware.

Figure 11.8 compares the IEEE 802.11i-based authentication with the IP-based authentication shown in Figure 11.2 (Section 11.2.2). The most significant change in the new authentication solution is that the AP participates in the authentication procedure. The terminal and the AP agree on the use of IEEE 802.1x authentication on association. The EAP authentication is then performed between the terminal and AP (1). The AP includes a RADIUS client module, which exchanges user authentication information with the back-end AS (2). The AS operates in a similar manner as in the IP-based solution. It queries the GSM HLR for authentication data and terminates the GSM SIM EAP protocol (3). After the EAP authentication has been completed successfully, the AP starts forwarding the user's IP data packets. Typically, the terminal will first proceed by obtaining an IP address with DHCP. Although the AC does not participate in the IEEE 802.1x authentication, it is still an integral part of the system. The AC implements support for IP-based authentication and includes important services such as DHCP server, RADIUS proxy server, local Web server, printer server, and location-based and access point management.

The Universal Mobile Telecommunications System (UMTS) is a new, global 3G mobile network standard. A UMTS subscriber will possess a UMTS Subscriber Identity Module (USIM), which is employed by the UMTS Authentication and Key Agreement (AKA) algorithms. Mobile operators can also use the operator wireless LAN solution as a complement to their UMTS networks. Because the USIM card is capable of running backward-compatible GSM/GPRS authentication algorithms, UMTS operators and subscribers can immediately use the GSM/GPRS-based operator wireless LAN solution. However, to take advantage of the enhanced security of UMTS subscriber authentication and UMTS key agreement, standardization activities are also in progress to define an EAP type for UMTS AKA [20]. Because operator wireless LAN authentication is based on EAP, no major changes to the architecture are required to support UMTS authentication. It is sufficient to use the UMTS AKA EAP type instead of the GSM SIM EAP type. The terminal and the AS, which are end points of the EAP negotiation, must include UMTS support and implement the UMTS AKA EAP type. Other operator wireless LAN components are not aware of the EAP type, and hence they need not be updated.

## 11.7 Conclusions

The increasing number of roaming users and the widespread use of broadband Internet services have created a strong demand for high-speed public Internet access with sufficient roaming capability. Wireless LAN systems offer high bandwidth but only modest IP roaming capability and global user management features. In this chapter, we have described a system that efficiently integrates wireless LAN access with the widely deployed GSM/GPRS roaming infrastructure. The architecture is designed to exploit GSM authentication, and SIM-based user management and billing mechanisms and combines them with public wireless LAN access. With the presented solution, the cellular operators can rapidly enter the growing broadband access market and utilize their existing subscriber management and roaming agreements. The operator wireless LAN system allows the cellular subscribers to use the same SIM (and user identity) for wireless LAN access. This gives the cellular operator a major competitive advantage compared with ISP operators who have neither a large mobile customer base nor a cellular-type roaming service.

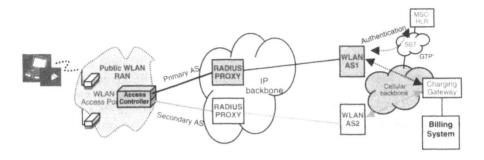

**Figure 11.7** Using RADIUS proxies for system redundancy.

The designed architecture combines cellular authentication with native IP access. This can be considered as the first step toward all-IP networks. The system proposes no changes to existing cellular network elements, which minimizes the standardization effort and enables rapid deployment. The reference system has been commercially implemented and successfully piloted by several mobile operators. GSM SIM-based wireless LAN authentication and accounting signaling has proved to be robust and scalable, offering a very attractive opportunity for mobile operators to extend mobility services to cover broadband indoor wireless networks.

## Acknowledgment

This chapter is an expanded version of the paper "Wireless LAN Access Network Architecture for Mobile Operators," that was published in *IEEE Communications Magazine* in November 2001.

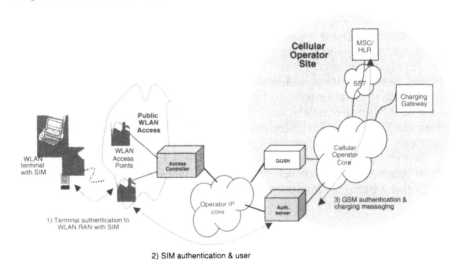

**Figure 11.8.** IEEE 802.11i-based solution.

# References

[1]  IEEE Std 802.11b, *Supplement to ANSI/IEEE Std 802.11, 1999 Edition, IEEE Standard for Wireless LAN Medium Access Control (MAC) and Physical Layer (PHY) specifications*, ISBN 0-7381-1812-5, 97 pp., January 2000.

[2]  Wireless Ethernet Compatibility Alliance, http://www.wirelessethernet.org/, June 2001.

[3]  IDC Personal Computer bulletin, "Worldwide Portable PC Forecast Update 1999-2003," document 21555, February 2000.

[4]  IEEE 802.1x, "Standards for Local Area and Metropolitan Area Networks: Standard for Port Based Network Access Control," March 2001.

[5]  IEEE 802.11i/D1.5, "Draft Supplement to Standard for Telecommunications and Information Exchange Between Systems—LAN/MAN Specific Requirements—Part 11: Wireless Medium Access Control (MAC) and Physical Layer (PHY) Specifications: Specification for Enhanced Security," August 2001 (work in progress).

[6]  M. Mouly and M. Pautet, *The GSM System for Mobile Communications*, Telecom Publishing, ISBN 0-9455-9215-9, 1992, 701 pp.

[7]  J. Hämäläinen, *Design of GSM High Speed Data Services*, Ph.D. Thesis, Tampere University of Technology, 104 pp., August 1996.

[8]  ETSI TS 101 393, Digital cellular telecommunications system (Phase 2+); General Packet Radio Service (GPRS); GPRS Charging, Version 7.6.0.

[9]  H. Haverinen, "NAAP: A User-to-Network Authentication Protocol," Submitted to Smartnet 2002, October 2001.

[10]  L. Blunk and J. Vollbrecht, "PPP Extensible Authentication Protocol (EAP)," IETF RFC2284, March 1998.

[11]  C. Rigney, A. Rubens, W. Simpson, and S. Willens, "Remote Authentication Dial In User Service (RADIUS)," IETF RFC2865, 2000.

[12]  C. Rigney, "RADIUS Accounting," IETF RFC2866, 2000.

[13]  B. Aboba and M. Beadles, "The Network Access Identifier," IETF RFC2486, 1999.

[14]  H. Krawczyk, M. Bellare, and R. Canetti, "HMAC: Keyed-Hashing for Message Authentication," RFC2104, 1997.

[15]  D. Fowler, *Virtual Private Networks—Making the Right Connection*, Morgan Kaufmann Publishers, Inc., ISBN 1-55860-575-4, 222 pp., 1999.

[16]  H. Haverinen, "EAP SIM Authentication (Version 1)," Internet-Draft draft-haverinen-pppext-eap-sim-01.txt, April 2001 (work in progress).

[17]  Telia HomeRun Business prices, http://www.homerun.telia.com/eng/business_pris.htm, 23 October 2001.

[18]  Sonera mBusiness Pricing List for Company Services, http://www.sonera.fi/mbusiness/hinnasto/index.html, 23 October 2001.

[19]  P. Huhtala, "Sonera wGate Wireless Internet", Merito Forum seminar on Access Networks End-to-End, 18–19 September 2001.

[20]  J. Arkko and H. Haverinen, "EAP AKA Authentication," Internet-Draft draft-arkko-pppext-eap-aka-00.txt, May 2001 (work in progress).

# Chapter 12

# From Wireless LANs to Wireless Network Systems:
## Applying Lessons from Cellular Networking to Enterprise Wireless Networking

Sandeep K. Singhal, Ph.D
*ReefEdge, Inc.*

Enterprises are extending their wired LAN infrastructure to support wireless connectivity, using radio standards such as IEEE 802.11b, Bluetooth, HiperLAN, and other emerging technologies. Wireless LANs give mobile users "anywhere, anytime" access to the LAN, and wireless connectivity offers lower installation and maintenance costs compared with traditional wired network infrastructure. However, to play a mission-critical role within the enterprise, wireless LANs must satisfy the security, manageability, interoperability, simplicity, seamless mobility, and application needs of network managers and end users.

In this chapter, we discuss the limitations of the wireless LAN standards as implemented within access points and client adapters. To bring security, manageability, usability, and performance to the wireless LAN, enterprises require additional network infrastructure. Borrowing from the design of cellular network infrastructure, the wireless LAN system creates a corporate-wide "Mobile Domain," a secure environment for mobile users within which they can access the information and resources they need to be more productive.

## 12.1 Introduction

Enterprises are extending their wired LAN infrastructure to support wireless connectivity, using radio standards such as IEEE 802.11b, Bluetooth, HiperLAN, and other emerging technologies. Wireless LANs give mobile users "anywhere, anytime" access to the LAN from clients such as laptops, handheld devices, or network-connected appliances. The wireless LAN infrastructure also provides connectivity for desktop PCs, offering lower installation and maintenance costs compared with traditional wired network infrastructure.

If wireless LANs are to play a mission-critical role within the enterprise, they must satisfy the disparate needs of network managers and end users. Network managers demand security, manageability, and interoperability; end users demand simplicity, seamless mobility, and applications that enhance their mobile experience.

To satisfy these requirements, enterprises cannot simply rely on the wireless LAN standards as implemented within access points and client adapters. These standards only define the radio link and how data is bridged between the wireless and wired LANs. To add security, manageability, usability, and performance to the wireless LAN, enterprises require additional network infrastructure.

In this chapter, we propose that enterprise wireless LAN deployments that include an in-building HLR/VLR are well positioned to support the needs of large-scale wireless systems. Moreover, by enabling additional services to access the HLR/VLR either through exposed APIs or directly, the wireless network system can evolve to address future challenges. With this infrastructure, a corporate-wide wireless network becomes a "Mobile Domain," a secure environment for mobile users within which they can access the information and resources they need to be more productive.

We first discuss the current limitations of wireless LANs within the corporate environment. After describing how a wireless network system infrastructure addresses these limitations, we consider the parallels between the in-building network system and a wide-area cellular network infrastructure.

## 12.2 Wireless LAN Connectivity

Enterprises of all sizes are deploying wireless LANs to extend or replace their existing wired LAN infrastructure. Wireless LAN deployments offer numerous advantages—mobility, convenience, and great flexibility. Wireless LANs deliver access to enterprise data and services from any location, at any time, and from any mobile device. They facilitate collaboration among employees and provide opportunities to deploy enhanced applications and services. Wireless LANs also offer reduced deployment and reconfiguration costs compared with wired LANs. As shown in Figure 12.1, these in-building wireless networks are comprised of Access Points (APs), which bridge data traffic between a wired network interface and a wireless radio interface. Client devices with wireless LAN adapters discover the nearby APs, establish a radio link to the nearest available AP, and transmit and receive data via the radio interface to the wired network.

Each AP in the wireless LAN provides reliable connectivity for multiple wireless clients, linking them to standard services such as DHCP, DNS, and IP routing located on the wired LAN. Beyond providing this connectivity and providing basic administrative and configuration interfaces through HTTP and SNMP, APs normally do not communicate with each other or with other network systems.

In early wireless LAN pilot deployments, enterprises experimented with wireless connectivity. In these trials, enterprises purchased several APs, placed them throughout the office to ensure coverage, configured them (to avoid channel overlap, for example), and connected them into one or more subnets on the existing wired network. Users access the network using their wireless LAN adapters.

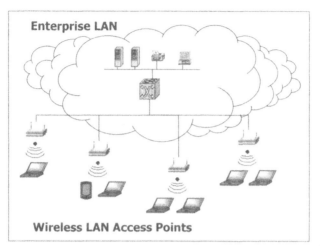

**Figure 12.1.** Traditional enterprise wireless LAN, consisting of Access Points that bridge traffic between the wireless and wireline networks.

## 12.3 Limitations of Wireless LANs

As enterprises migrate toward full deployment of wireless LAN infrastructure, they are encountering the limitations posed by basic wireless connectivity. APs alone cannot deliver the necessary capabilities to support a scalable, mission-critical wireless deployment throughout a building or campus.

### 12.3.1 Security

The wireless environment introduces numerous security challenges. Possible security threats include unauthorized use of the network, eavesdropping on transmitted data traffic, and denial of service attacks. Although these threats are present in traditional LANs, the wireless environment significantly exacerbates the situation for the following reasons:

- Wireless LAN range and signal propagation are largely uncontrollable, so potential intruders on the LAN need not be physically located within an enterprise's premises;

- 802.11b networks offer a range of at least 100 m, often extending coverage into the parking lot and maybe even to nearby streets [3]. This concern is particularly relevant in shared office buildings.

- With the emergence of handheld devices capable of communicating over 802.11b and Bluetooth networks, the inability to monitor user actions is of even greater concern. Even within an enterprise, a wireless LAN intruder can operate inconspicuously because he does not need a physical connection to the network;

- Guests and visitors increasingly expect enterprises to offer Internet access, in the same way that they ask to borrow the telephone or fax machine today.

Unfortunately, as layer 2 technologies, wireless LAN standards do not define the necessary mechanisms to address these concerns within an enterprise setting. Instead, the standards only provide for link-level encryption for traffic transmitted over the air; this encryption is often unsuitable for enterprise-level security. The vulnerabilities [1, 2, 4] in the Wired Equivalent Privacy (WEP) algorithms in 802.11b are well documented and include attacks on device authentication protocol, data analysis attacks to determine the shared encryption key, and data integrity attacks allowing modification of data transmissions or clandestine traffic insertion.[1]

To satisfy the security requirements of an enterprise wireless LAN, additional network infrastructure is required to protect the network from malicious or unapproved usage. This network infrastructure must satisfy three security requirements:

- **Authenticate all wireless LAN users.** Before gaining network access through the wireless LAN, all users must identify themselves. At a minimum, this authentication requires a username and password, but alternative identification technologies such as Secure ID and biometrics may also be appropriate in some environments;

- **Control network use by wireless LAN users.** With different user populations (employees, guests, contractors, and so forth) gaining network access through the wireless LAN, network infrastructure must control how those users are able to use the network. In particular, the infrastructure must enforce class-based access control policies that may restrict access to particular hosts, networks, ports, and protocols;

- **Ensure adequate data privacy.** Given the ineffectiveness of link-layer encryption, the network infrastructure must ensure that encryption policies can be both enforced and implemented effectively. In particular, enterprises already rely on VPN technologies such as IPsec to provide data encryption over untrusted links. However, traditional VPN solutions do not scale to support enterprise-wide use, and they are expensive. The wireless LAN infrastructure should support the efficient and cost-effective deployment of IPsec for access to protected servers, without mandating its use for accessing Internet hosts (which already rely on higher-level encryption such as SSL).

---

[1] The insecurity of WEP eventually will be addressed by introducing alternative encryption techniques (such as AES, as being defined in the 802.11i Task Group). However, enterprises will continue to demand software-based encryption such as IPsec, both because it is compatible with existing encryption techniques deployed in the enterprise and because security flaws can be more readily addressed through software upgrades.

### 12.3.2 Mobility

Wireless LANs do not enable seamless mobility throughout the wireless network. The wireless LAN standards only define how an AP bridges wireless traffic with the local LAN segment. This layer 2 bridging restricts users, who must communicate through multiple APs, each possibly connected to different LAN subnets. The mobility problem arises because the client must change IP addresses as it roams about the wireless network. IP addresses enable the Internet to route traffic that is destined for a particular host. To enable this routing, therefore, each IP address must be permanently associated with a particular LAN subnet. To change subnets, a client must first relinquish its previous IP address and obtain an IP address that is linked to the new subnet.

Unfortunately, for a client to change its IP address, it must also terminate all ongoing network connections. At best, assuming the application is carefully engineered to reestablish a new connection, this causes reduced data throughput, noticeable breaks in real-time multimedia streaming and presentation, or the need to restart an operation. At worst, an application may deliver incorrect results to the user, stall, or crash. Even stationary users may encounter these subnet roaming problems. When located in areas with weak wireless coverage or overlapping coverage, the client's wireless adapter can automatically switch between APs according to changes in ambient radio interference.

To allow the wireless APs to cooperate, enabling seamless mobility, network infrastructure must ensure that each client receives an IP address that remains constant regardless of the client's location in the network. In this environment, the enhanced network infrastructure must handle packet routing based on its knowledge of the user's physical location rather than being simply based on the IP address assigned to that user.

### 12.3.3 Quality of Service

Most wired LANs today run at 100 Mbit/s, with switched LANs delivering a dedicated data rate of 100 Mbit/s to each user; many enterprises are moving toward 1 Gbit/s and even 10 Gbit/s over their wired networks. Given this bandwidth, users do not give a thought to running multimedia, voice, streaming video, and thin GUI applications (e.g., SAP, Oracle ERP systems) over these networks to their desktop. On the other hand, IEEE 802.11b networks are rated at 11 Mbit/s, but the actual throughput typically does not exceed 7 Mbit/s because of radio collisions in a multiuser wireless LAN environment. The wireless LAN bandwidth is shared among all users utilizing the AP, so an individual user's effective bandwidth can average a meager 300 kbit/s—and worse if another user happens to begin a large MP3 download! Newer wireless LAN technologies that promise 50 Mbit/s improve the situation, but the per-user capacity is still far below that of a wired network. The situation is equivalent to dialing into a high-speed network over a low-quality analog cell phone, where "cross talk" limits your ability to communicate effectively. The obvious result is that modern desktop applications may not run effectively over wireless LAN networks that cannot deliver the necessary bandwidth.

Wireless LANs require *effective Quality-of-Service (QoS) management* to monitor and control how much bandwidth is used by each network user. A policy-driven approach can ensure that the limited bandwidth is allocated fairly among the active users. It can ensure that guest users do not block bandwidth availability from employees, and it can guarantee that key bandwidth-intensive applications receive the bandwidth they require. To support enterprise wireless LAN deployments, additional network infrastructure is required to centrally define, manage, and enforce QoS policies for all wireless APs. These policies divide the available wireless bandwidth among user groups, applications and protocols, and traffic classes in accordance with enterprise needs.

### 12.3.4 Management and Monitoring

An enterprise wireless LAN deployment may involve hundreds of APs from multiple vendors using different wireless technologies. Each AP typically provides administrative interfaces (exposed to Web browsers through HTTP or to network management systems through SNMP), but these interfaces offer limited value in an enterprise environment. The Web interfaces only enable a network administrator to configure an individual AP. Traditional network management tools provide limited monitoring and event logging capabilities, but they do not support full management of the policy and control attributes that are particular to the wireless LAN environment.

Large-scale wireless LAN management today is both complex and error prone. The administrator must identify and inventory all APs, configure and update them individually, and define monitoring processes for the entire collection. Clearly, as in the wired network—in which centralized administration and monitoring are standard—the IT manager must gain control over all elements of the wireless infrastructure. At the same time, the IT manager must integrate the wireless LAN with existing management infrastructure in the wired network. To support an enterprise-wide wireless LAN, additional network infrastructure is required to provide centralized configuration, management, and monitoring capabilities for all elements of the wireless network. This management infrastructure must provide:

- **Network configuration.** Define global wireless network parameters such as the SSID, WEP keys, and location of network services;
- **Policy definition.** Define security, access control, and QoS policies;
- **Performance monitoring.** Access real-time and historical records of traffic levels, with information about traffic destinations, protocols, and usage by user/group;
- **Authentication and directory integration.** Link the wireless network with existing authentication services and directory services for identifying users and groups, retrieving access control and policy information, and linking to enterprise asset information;
- **Enterprise system integration.** Link the wireless network with standard services such as event management, system logging, and intrusion detection.

### 12.3.5 Enhanced Application Support

Because wireless LAN technologies are primarily network technologies, their only service to applications involves delivering connectivity to wireless users. Through a wireless LAN, users simply gain wireless access to the same applications that they can access over the wireline network. However, wireless users are different from wireline users. First, they are more likely to be mobile, working away from their desks. These users may operate in unfamiliar surroundings, meaning that they need easy access to the resources located physically nearby. These users may not have access to the information and data resources that are otherwise available at their desks, so they need remote access to their desktop computers and applications. Second, they are increasingly likely to use new handheld devices—PDAs, tablets, and even cell phones—with short-range wireless capabilities. These devices offer limited user interface capabilities, but within an enterprise wireless environment they may be appropriate tools for rapidly accessing and manipulating information.

The wireless LAN deployment offers the opportunity to deliver a new generation of application services that improve wireless user productivity. Within the wireless LAN, applications can offer single sign-on, personalize services according to the user's location and device type, and offer efficient remote access to desktop data and enterprise services. To deliver these new applications, the enterprise requires new network infrastructure that allows applications to become aware of the wireless network. This infrastructure must deliver secure APIs exposing user, device, and location information and allowing applications to retrieve the information required to personalize the user experience. The wireless network infrastructure must also enhance intranet portals with user device and location information.

## 12.4 Wireless Network Systems

As we have seen, wireless APs alone cannot meet the needs of an enterprise-wide wireless LAN deployment: seamless mobility, comprehensive security, QoS enforcement, management and monitoring, and enhanced application support. Even when a particular AP does provide some of these features, it is usually non-interoperable, because wireless LAN standards only define the correct behavior of a network bridge. The wireless LAN requires network infrastructure that converts the APs into an integrated wireless network. Instead of installing individual APs, enterprises are moving to complete wireless LAN systems. An enterprise *wireless network system* combines the wireless LAN APs with the infrastructure necessary to address the unique requirements of a wireless LAN deployment.

### 12.4.1 Wireless System Infrastructure

The wireless LAN network infrastructure is a distributed system that includes a collection of edge services and a network coordination server. As shown in Figure 12.2, these components interconnect the wireless APs together to form a single communication, security, and management infrastructure.

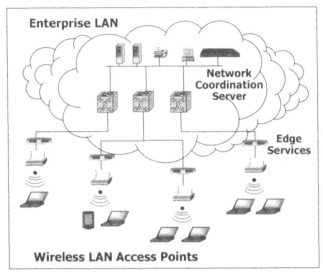

**Figure 12.2.** An enterprise wireless LAN system, including edge services and a network coordination server that stitch together the wireless APs into a single communication, security, and management infrastructure.

### 12.4.2 Edge Services

Edge services are hosted within appliances that are installed behind the existing wireless APs. These edge services intercept and process all traffic being exchanged between the wireless and wireline networks to enforce the configuration, security, and QoS policies of the wireless system. By operating on all wireless network traffic, the edge services eliminate the need to install client-side software to enforce wireless system policies. The edge services perform the following operations:

- Control access to the wireless network by enforcing user authentication, filtering disallowed traffic, and enabling data encryption;
- Enable seamless communications by enabling global address allocation, session persistence, and session mobility across wireless APs;
- Improve performance of the wireless network by enforcing QoS guarantees on a per-user or per-user group basis; enabling bandwidth reservations; and prioritizing network traffic according to its host, port, or application.

The decentralized operation of edge services delivers several advantages within the wireless network system. First, edge services reduce the processing overhead by operating on packets along their normal routing path between wireless and wireline networks. The alternative, a central server, requires that packets be artificially routed to that server; this introduces additional latency. Second, edge services avoid the single bottleneck and single point of failure that a central server represents. Third, edge services ease network deployment and evolution. New APs (with correspondent edge services) can be installed, and the AP automatically becomes part of the wireless network system.

Edge services operate in conjunction with the network coordination server to enable the wireless network system to operate cohesively. Edge services receive network configuration, security policies, and QoS guidelines from the network coordination server. They use this information to process wireless packets in accordance with the system policies. Each edge services appliance is also responsible for ensuring proper configuration of the wireless AP(s) under its control. Edge services also deliver configuration, performance, and traffic reports for logging, accounting, and monitoring purposes.

### 12.4.3 Network Coordination Server

The network coordination server brings together the wireless edge services under a single point of administrative control. The server does not process the client's wireless traffic, because the edge services hold that processing responsibility. Instead, the server coordinates the behavior of the edge services to ensure consistent operation throughout the wireless network system.

The network coordination server plays several roles in the wireless network system:

- Hold the wireless network configuration, security policies, and QoS guidelines for the wireless network system;
- Discover the installed edge services and provide them with up-to-date configuration and policy information;
- Record dynamic information about active user communications and coordinate the connection handoff as the user roams between APs spanning different edge service appliances;
- Monitor the status of edge services and wireless users;
- Integrate the wireless network system with the existing managed LAN infrastructure and existing enterprise information systems, including authentication services, logging services, audit and intrusion detection services, directory services, and network management systems;
- Expose application and portal APIs exposing information about the identity and location of users, devices, and resources.

## 12.5 Lessons from Cellular Infrastructure

The requirements of the enterprise wireless LAN bear a striking resemblance to those of a cellular infrastructure for wide-area wireless communications. To gain insight into the design of an in-building wireless network system, it is therefore instructive to consider the operation of a cellular infrastructure, in particular, a packet-oriented General Packet Radio Service (GPRS) environment.

### 12.5.1 Overview of Packet-Oriented Cellular Communications

Like an in-building wireless network, a cellular network is comprised of wireless-enabled client devices (cell phones) and various base stations supporting wireless signaling standards such as GSM, CDMA, and TDMA. Although these components are the most visible aspects of a cellular network, the network itself is only useful when those base stations are tied together to enable handoff and roaming, identify subscribers and control their access to the wireless network, and deliver presence and location information to mobile applications and services. This infrastructure underlying the cellular network is the most performance-critical—and usually the most expensive—part of the system. Figure 12.3 illustrates some of the infrastructure underlying a packet-based cellular network.

A *Radio Network Controller* (RNC) links to a collection of base stations and coordinates all traffic sent through those base stations. It ensures that all wireless devices are properly authenticated and enforces access policies. The RNC authenticate users and receive policies by coordinating with a *Serving GPRS Support Node* (SGSN) server located within the carrier's network. The SGSN is in charge of packet routing and connecting the cellular network to other networks. Underlying the SGSN are two important databases, the *Home Location Register* (HLR) and *Visiting Location Register* (VLR). The HLR records static information about the registered cellular users and devices, their respective access rights, QoS guarantees, charging information, and other preferences. The VLR records dynamic information about the active cellular users and devices (which may be users from the local HLR or from the HLR of another carrier through a roaming agreement). The VLR records the current location of a user/device (i.e., the appropriate RNC), traffic information, active call status, and so on.

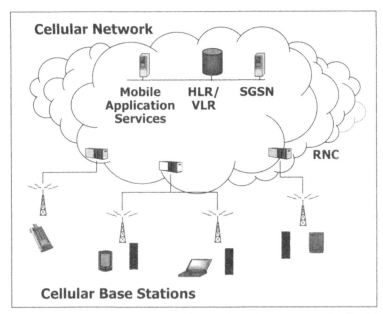

**Figure 12.3.** Packet-based cellular infrastructure incorporates Radio Network Controllers (RNC) that manage clusters of base stations, a Serving GPRS Support Node (SGSN), Home Location Register (HLR) and Visiting Location Register (VLR) databases, and mobile application services.

The HLR/VLR form the heart of the cellular network. These high-performance systems support thousands of queries and updates per second. In response to user behavior, the SGSN communicates with the HLR/VLR to obtain and record current state information about the network. Although the SGSN is the primary user of the HLR/VLR, it is by far not the only service that relies on these databases. The cellular network typically includes other servers that communicate with the HLR/VLR (through exposed APIs and protocols) to support Mobile Internet services, location-based services, short message service, prepaid billing, and other capabilities.

### 12.5.2 An In-Building HLR/VLR

The parallels between an in-building wireless network system and the cellular environment are quite clear: The edge services parallel the operation of the cellular RNC, whereas the network coordination server parallels the operation of the cellular SGSN. The implementation of an in-building network coordination server requires the deployment of a high-performance data repository—an in-building HLR/VLR targeted for the wireless LAN, as shown in Figure 12.4. To support the mobility, security, QoS, management and monitoring, and application services of the network coordination server, the in-building HLR/VLR includes the following fields:

- User identity, according to which user has authenticated from a particular device;
- Device identity, consisting of the device's MAC address and encryption keys;
- Addressing information, consisting of the device's assigned IP address;
- Routing and location information, identifying the edge services appliance(s) through which the client is currently communicating;
- Connection information, recording active data connections associated with this client.

These fields link to other information tables, including access control policies and QoS policies.

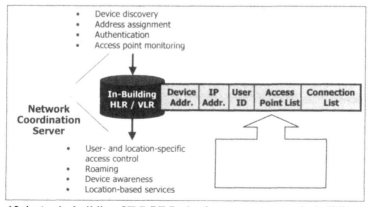

**Figure 12.4.** An in-building HLR/VLR database supports an in-building wireless network system, borrowing from the design of cellular environments.

## 12.6 Conclusions

Enterprise wireless LAN deployments pose new challenges to IT managers. Individual APs cannot deliver the mobility, security, QoS, management and monitoring, and application capabilities required in large-scale wireless LAN deployments. To address these concerns, the wireless LAN requires a wireless network system consisting of edge services, a network coordination server, and an in-building HLR/VLR.

ReefEdge's first product, the ReefEdge Connect System, delivers comprehensive security, performance, mobility, manageability, and scalability without a major investment in hardware, network reconfiguration, or software development. Connect works with all of today's wireless LAN standards and is positioned to capitalize on future developments in wireless technology. The system consists of the ReefEdge Connect Server, which allows centralized management of the system, and ReefEdge Connect Bridges, which act as "microfirewalls" at the edge of the network, implementing the system's mobility, security, and management features. The Connect System works with all mobile devices and does not require the installation of new client software. The Connect System is easily layered onto an existing deployment of wireless APs. The solution provides controls necessary to manage users, applications, and services within the network.

By marrying the network connectivity layer and the application layer of the system, ReefEdge delivers fine-grained location-based information, access, and resource control—a level of control that did not exist before the advent of the ReefEdge solution. Our solution also links networking capabilities to an application platform. This functionality allows system administrators to securely deploy and manage mobile devices and safely integrate them into the enterprise-computing environment. The knowledge of user location and identity is used to integrate with existing enterprise applications to deliver personalized, location-aware, and collaborative services.

The enterprise of the future is one in which every employee will be a mobile user and in which every laptop and handheld device will be wireless. The mobile workforce will collaborate more easily, be more productive within flexible office configurations, and travel easily between buildings in a campus environment or across office sites worldwide. ReefEdge products allow the enterprise to deploy laptops, PDAs, and other devices as "first-class" computing tools today. As a result, the enterprise can harness the increased productivity, efficiency, and power of a truly mobile workforce.

## References

[1]    W. A. Arbaugh, N. Shankar, and Y. C. Wan, "Your 802.11b Network has No Clothes,"        March        2001.        Available        from http://www.cs.umd.edu/~waa/wireless.pdf

[2] N. Borisov, I. Goldberg, and D. Wagner, "Intercepting Mobile Communications: The Insecurity of 802.11," *Proceedings of the ACM SIGMOBILE Seventh Annual International Conference on Mobile Computing and Networking*, July 2001. Available from http://www.isaac.cs.berkeley.edu/isaac/mobicom.pdf

[3] J. Leyden, "War Driving—The Latest Hacker Fad," *The Register*, 29 March 2001. Available from http://www.theregisgter.co.uk/content/8/17976.html

[4] J. R. Walker, "Unsafe At Any Key Size: An Analysis of the WEP Encapsulation," Intel Corporation (Document IEEE 802.11-00/362), October 20, 2000. Available from http://grouper.ieee.org/groups/802/11/Documents/DocumentHolder/0-362.zip

# Chapter 13

# The Bluetooth Basics

Mike Sheppard
*Bluetooth SIG Associate Member*

Bluetooth is a short-range wireless communication standard that enables personal area networking among a wide variety of electronic devices, ranging from laptops to cell phones, computers to printers, personal digital assistants to wireless headsets, and many other devices and applications. This chapter discusses the evolution of Bluetooth and the essential technologies encompassing the standard and provides insights on the future success of the standard.

## 13.1 Introduction

Bluetooth is a global standard that:

- Eliminates wires and cables between both stationary and mobile devices;
- Facilitates both data and voice communication;
- Offers the possibility of ad hoc networks and delivers the ultimate synchronicity between all personal devices.

Bluetooth is a wireless, personal area network specification defining a short-range, low-power connectivity solution for voice and data communications. The Bluetooth wireless technology includes a universal radio frequency interface between portable electronic devices in short-range, ad hoc networks and comprises hardware, software, and interoperability requirements. Bluetooth wireless technology enables devices to communicate without regard to cable and connectors. The technology has been adopted by all major players in the telecommunications, computer, and home entertainment industries and has extended in other diverse areas such as the automotive industry, health care, automation, etc., almost all sectors of the economy.

### 13.1.1 What's with the Name?

Harald Bluetooth was a Viking and king of Denmark between 940 and 981. One of his skills was to make people talk to each other. During his rule, Denmark and Norway were Christianized and united. Today's Bluetooth wireless technology enables people to talk to each other, but this time by means of a low-cost, short-range radio link. In the Danish town of Jelling, Harald Bluetooth raised an enormous rune stone that still stands in its original position. It has the following runic inscription, adorned with an image of Christ: "King Harald raised this monument to the memory of Gorm his father and Thyre his mother, that (same) Harald which won all Denmark and Norway and made the Danes Christian." In September 1999 a new stone was raised outside of Ericsson Mobile Communications in Lund, Sweden, this time to the memory of Harald Bluetooth.

### 13.1.2 The First Steps

The idea that resulted in the Bluetooth wireless technology was born in 1994 when Ericsson Mobile Communications initiated a study to explore a low-cost wireless cable replacement (then called MC link). The purpose of MC link was to eliminate the need for cables between mobile phones and their accessories. As the project investigated the use of cheap, short-range radio, they quickly discovered a variety of unexpected applications. A year later, Ericsson designers started developing the transceiver and the true potential of the technology started to crystallize. But beyond unleashing devices by replacing cables, the radio technology showed possibilities for becoming a universal bridge to existing data networks, a peripheral interface, and a mechanism to form small, private ad hoc groupings of connected devices away from fixed network infrastructures.

## 13.2 The Bluetooth Special Interest Group

By early 1998 Ericsson recruited four strategic associates, Intel, IBM, Nokia, and Toshiba, and the Bluetooth Special Interest Group (SIG) was formed. The assignment of the SIG originally was to monitor the technical development of short-range radio and to create an open global standard, thus preventing the technology from becoming the property of a single company. The SIG developed and released the Bluetooth Core and Bluetooth Profile standard specification (version 1.0A) in late 1999. Today, the Bluetooth SIG includes promoter companies 3Com, Ericsson, IBM, Intel, Lucent, Microsoft, Motorola, Nokia, Toshiba, and thousands of adopter/associate member companies. Further development of the Bluetooth specification is still one of the main issues for the SIG. Other important tasks are interoperability requirements, frequency band harmonization, and promotion of the technology.

### 13.2.1 Interoperability

From the very start, one of the main goals for the SIG has been to include a regulatory framework in the Bluetooth specification that will guarantee full interoperability between different devices from various manufacturers—as long as they share the same Profile. Whereas the usage models describe applications and intended devices, the Profiles specify how to use the Bluetooth protocol stack for an interoperable solution. Each Profile states the methods to reduce options and set parameters in the base standard and how to use procedures from several base standards. A common user experience is also defined. For example, a computer mouse does not need to communicate with a headset, and so they are built to comply with different profiles.

The Profiles are a part of the Bluetooth specification, and all devices must be tested against one or more of the Profiles to fulfil the Bluetooth certification requirements. The number of Profiles will continue to grow as new Bluetooth applications arise.

### 13.2.2 Compliance

The Bluetooth Qualification Program guarantees global interoperability between devices regardless of the vendor and regardless of the country in which they are used. During the test procedure that all devices must pass, it must be verified that they meet all requirements regarding:

* Radio link quality;
* Lower-layer protocols;
* Profiles;
* Information to end users.

All qualified devices are listed at the SIG official website.

### 13.2.3 Usage Models

The profiles defined in the first version of the Bluetooth specification mainly address usage models concerning the telecommunication and computing industries. Three examples include the "Internet Bridge," the "Ultimate Headset," and the "Automatic Synchronizer":

* An **Internet Bridge** giving constant access to the Internet is a useful and time-saving feature, especially when the bandwidth of mobile phones is increasing rapidly. Bluetooth wireless technology allows you surf the Internet without any cable connections wherever you are, either by using a computer or by using the phone itself. When close to a wire-bound connection point, your mobile computer or handheld device can also connect directly to the landline, but still without cables;

* The **Ultimate Headset** allows you to use your mobile phone even if it is placed in a briefcase, thereby always keeping your hands free for more important tasks when you're at the office or in your car;

* **Automatic synchronization** of calendars, address books, etc. is a feature long-awaited by many people. Simply by entering your office, the calendar in your phone or Personal Digital Assistant (PDA) will be automatically updated to agree with the one in your desktop PC, or vice versa. Phone numbers and addresses will always be correct in all your portable devices without docking through cables or infrared.

## 13.3 Bluetooth Products

Many companies have declared that Bluetooth wireless technology will be incorporated into their products, especially when components become cheaper. In a forecast made by Cahners In-Stat Group in July 2001, the product availability during the next couple of years was defined as three waves:

The **first wave** is believed to occur sometime in 2002 and will include products such as:

- Adapters for mobile phones and adapters/PC cards for notebooks/PCs;
- High-end mobile phones and notebooks with integrated Bluetooth communication for business users;
- Bluetooth headsets;
- Cordless phones, handheld PCs, and PDAs.

The **second wave** will overlap the first in many aspects. Expected products include:

- PCs with Bluetooth circuitry on motherboards;
- Printers, fax machines, digital cameras, products for industrial, medical, and vertical industries;
- Bluetooth automotive solutions involving hands-free mobile phone usage with regular mobile phones.

The **third wave** will include:

- Low-cost mobile phones and lower-cost portable devices/desktop PCs.

## 13.4 Why Bluetooth Wireless Technology?

In phase with the IT boom, the mobility among people has constantly grown and wireless technologies for voice and data have evolved rapidly during the past few years. Countless electronic devices for home, personal, and business use have been presented to the market during recent years but no widespread technology to address the needs of connecting personal devices in Personal Area Networks (PANs). The demand for a system that can easily connect devices for transfer of data and voice over a short distances without cables grew stronger.

Bluetooth wireless technology fills this important communication need, with its ability to communicate both voice and data wirelessly, using a standard low-power, low-cost technology that can be integrated in all devices and thus enable total mobility. The price will be low, thereby resulting in mass production. The more units around, the more benefits for the customer.

### 13.4.1 Bluetooth Characteristics

Bluetooth provides a reliable, wireless packet-based, two-way communication medium allowing heterogeneous devices to communicate in ad hoc PANs. The packets can carry both data and voice from one Bluetooth-enabled device to another over a radio frequency air interface.

In the development of Bluetooth wireless technology, the SIG follows a fundamental design principle: Keep it simple. Simple designs promote simple implementations. Simple implementations can be small, cheap, and thrifty. Requirements regarding size, power, and price were set to move toward Bluetooth ubiquity. The radio unit should be so small and consume so little power that it could enable portable devices of all kinds; from pens to cameras, cell phones to earphones (Fig. 13.1):

- **Low Power:** Preliminary data sheets reported that a Bluetooth radio in standby mode requires 10 µA. While transmitting or receiving, the maximum current required is 50 mA;
- **Small Size:** The Bluetooth SIG projects that silicon chips will average 9 mm². Manufacturers currently offer multichip and single-chip solutions. For example, the Anoto™ pen consists of three main parts: the digital camera, an advanced image processing unit, and a Bluetooth radio transceiver;
- **Low Price:** The incremental cost goal for Bluetooth is $5 per device.

### 13.4.2 The Technology

The Bluetooth specification defines a short (around 10 m)- or optionally a medium (around 100 m)-range radio link capable of voice or data transmission to a maximum capacity of 720 kbit/s per channel. Radio frequency operation is in the unlicensed Industrial, Scientific, and Medical (ISM) band at 2.4 GHz, using a spread spectrum, frequency hopping, full-duplex signal at up to 1,600 hops/s. The signal hops among 79 frequencies at 1-MHz intervals to give a high degree of interference immunity. RF output is specified as 0 dBm (1 mW) in the 10-m-range version and −30 to +20 dBm (100 mW) in the longer-range version. When producing the radio specification, high emphasis was placed on making a design enabling single-chip implementation in CMOS circuits, thereby reducing cost, power consumption, and the chip size required for implementation in mobile devices.

Low power consumption enables battery-driven applications

Low profile allows units to be embedded in small appliances

Low cost enables ubiquity (devices everywhere)

**Figure 13.1.** Bluetooth characteristics.

### 13.4.3 Band Allocation

Bluetooth operates on the 2.4-GHz license-free ISM band (Table 13.1) that enjoys worldwide support and is open to any radio system with slight frequency variation from country to country. Frequency of 2.45–2.4835 GHz is used in the United States and Europe. The SIG is working diligently to promote worldwide frequency harmonization. For the majority of countries, the operational band is 79 individual carrier frequencies at 1-MHz intervals. Table 3.2 shows the radio parameters used in Bluetooth.

### 13.4.4 Voice

Each voice channel supports a 64 kbit/s synchronous voice channel in each direction. Up to three simultaneous synchronous voice channels can be used. Alternatively, a channel that simultaneously supports asynchronous data and synchronous voice can be used.

### 13.4.5 Data

The asynchronous data channel can support maximal 723.2 kbit/s asymmetric (and still up to 57.6 kbit/s in the return direction) or 433.9 kbit/s symmetric.

## 13.5 Network Architecture

Bluetooth units that come within range of each other can set up ad hoc, point-to-point and/or point-to-multipoint connections. Units can dynamically be added to or disconnected from the network. Two or more Bluetooth units that share a channel form a Piconet. Several piconets can be established and linked together in ad hoc Scatternets to allow communication and data exchange in flexible configurations (Fig. 13.2). If several other Piconets are within range, they each work independently and each have access to full bandwidth. Each Piconet is established by a different frequency hopping channel. All users participating on the same Piconet are synchronized to this channel. Unlike infrared devices, Bluetooth units are not limited to line-of-sight communication.

### 13.5.1 Scatternet

To regulate traffic on the channel, one of the participating units becomes a master of the Piconet, and all other units become slaves. With the current Bluetooth specification, up to seven slaves can actively communicate with one master. However, there can be an almost unlimited number of units virtually attached to a master being able to start communication instantly on a first come, first serve basis. Slaves can participate in different Piconets, and a master of one Piconet can be the slave in another, forming a Scatternet. Up to 10 Piconets within range can form a Scatternet, with minimum collisions.

### Table 13.1. Band allocation.

| Region | Regulatory Range | Channels |
|---|---|---|
| United States, Europe | 2.4–2.4835 GHz | 79 |
| Spain | 2.445–2.475 GHz | 23 |
| Japan | 2.400–2.4835 GHz | 79 |

### Table 13.2 Bluetooth radio parameters.

| | |
|---|---|
| Modulation | GFSK |
| Peak data rate | 1 Mbit/s |
| RF bandwidth | 220 kHz (–3 dB), 1 MHz (–20 dB) |
| RF band | 2.4-GHz ISM band |
| RF carriers | 23 (Spain) or 79 (US, Europe, Japan) |
| Carrier spacing | 1 MHz |
| Peak transmit power | <20 dBm |

### 13.5.2 Addressing

A master can share an asynchronous channel in a Piconet with up to seven simultaneously active slaves using the 3-bit active member address (AM_ADDR). By swapping active and inactive (parked) slaves respectively in the Piconet, 255 slaves can be virtually connected using the 8-bit parked member address (PM_ADDR). A slave device can participate in an active session, become parked, and participate again within 2 ms. To park even more slaves, the 48-bit Bluetooth device address (BD_ADDR) can be used. There is no limitation to the number of slaves that can be parked.

### 13.5.3 Security

Because radio signals can be easily intercepted, Bluetooth devices have built-in security to prevent eavesdropping or falsifying the origin of messages (spoofing). Bluetooth applications may choose from several levels of error-correction encoding techniques to facilitate reliable communication. The technology also provides for several levels of secure communication by stipulating protocols and procedures for authentication, authorization, and encryption (at the hardware level as well as the software level). Because of its strong security features and interface management procedures, Bluetooth enables concurrent networks in the same geographic space. In fact, as illustrated in Figure 13.2, Bluetooth devices can participate in more than one network at a time.

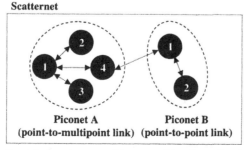

**Figure 13.2.** A Scatternet showing an ad hoc combination of point-to-multipoint and point-to-point Piconets.

The main security features are:

- A challenge-response routine—for authentication (which prevents spoofing) and unwanted access to critical data and functions;
- Stream cipher—for encryption, which prevents eavesdropping and maintains link privacy;
- Session key generation—session keys can be changed at any time during a connection.

Three entities are used in the security algorithms:

- The 48-bit Bluetooth device address (BD_ADDR) is a public entity unique for each device and can be obtained through the inquiry procedure;
- The 128-bit private user key is a secret entity that is derived during initialization and is never disclosed;
- A 128-bit random number is derived from a pseudorandom process in the Bluetooth unit, generating a different number for each new transaction.

In addition to these link-level functions, frequency hopping and the limited transmission range also help to prevent eavesdropping.

## 13.6 Hardware Architecture

Bluetooth-enabled devices consist of four basic components. The radio (with antenna) and link controller are hardware entities. The stack and application are software entities. Here we will discuss the hardware entities in more detail.

The Bluetooth hardware consists of an analog radio part and a digital part (the Host Controller). The Host Controller has a hardware digital signal processing part (called Link Controller), a CPU core, and interfaces to the host environment (Fig. 13.3). The Link Controller consists of hardware that performs baseband processing and physical layer protocols (e.g., automatic repeat request) and FEC coding. The function of the Link Controller includes asynchronous transfers, synchronous transfers, audio coding, and encryption. The CPU core allows the Bluetooth module to handle Inquiries and filter Page requests without involving the host device. The Host Controller can be programmed to answer certain Page messages and authenticate remote links.

The Link Manager (LM) software runs on the CPU core. The LM discovers other LMs and communicates with them via the Link Manager Protocol (LMP) to perform its service provider role and to use the services of the underlying Link Controller.

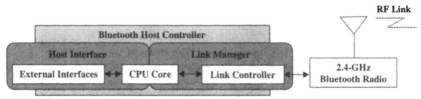

**Figure 13.3.** Bluetooth hardware components.

## 13.7 Software Architecture

The Bluetooth protocol stack and source application are software entities. Where the software resides depends on the current application. Computational hosts locate most of the software stack and application in the main memory. These hosts must adapt the Bluetooth architecture to a bus or serial port transport. If the application is embedded, the application locates the software components within a microcontroller device along side the link controller and radio—in hardware.

The Bluetooth protocols are the LMP, baseband, and audio blocks in Figure 13.4. In order to make different hardware implementations compatible, Bluetooth devices use the Host Controller Interface (HCI) as a common interface between the Bluetooth host (e.g., a portable PC) and the Bluetooth core.

Higher-level protocols like the Service Discovery Protocol (SDP), RFCOMM (emulating a serial port connection), and the Telephony Control Service (TCS) are interfaced to baseband services via the Logical Link Control and Adaptation Protocol (L2CAP). Among the issues L2CAP takes care of are segmentation and reassembly to allow larger data packets to be carried over a Bluetooth baseband connection. The SDP allows applications to find out about available services and their characteristics when devices are moved or switched off.

### 13.7.1 Bluetooth Component Map

A closer look at the Bluetooth components reveals how the protocols are arranged. The physical layer corresponds to the radio (in hardware). The link controller hardware implements low-level link control using the LMP and the basic Bluetooth baseband functionality. In addition, the Bluetooth device offers a HCI allowing upper levels implemented in software to initiate and respond to lower-level hardware services and events. It is also seen that the specifications adopt some pre-existing application content standards (such as vCard, vCalendar, and others in the IrMC family) as well as emerging wireless application environment protocols. A Bluetooth application may access audio directly (from the hardware), may control the link controller baseband directly (via the HCI), and may transmit and receive data (via the HCI) through intermediate software layers supported by the L2CAP.

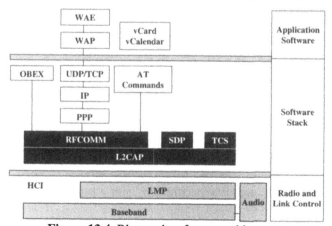

**Figure 13.4.** Bluetooth software architecture.

## 13.8 Profiles and Usage Models

The Bluetooth SIG currently defines and supports 14 profiles, although several others are under development. The five general profiles, on which most others depend, are:

- General Access;
- Serial Port;
- Service Discovery;
- Generic Object Exchange;
- TCS Binary.

Nine other concrete profiles depend on these general profiles. At the top of the dependency tree is the Generic Access Profile that defines services and procedures common to all other operational contexts.

When involved in radio communication according to a Bluetooth profile, each device assumes one of the roles specified in a profile, such as gateway or voice terminal, network access point or data terminal, headset or audio server, Object Exchange (OBEX) client or server. Usage models describe the prime Bluetooth applications and intended devices, whereas profiles specify how the interoperability solution for the functions described in the usage models should be provided with Bluetooth protocols. A brief description of the major profiles follows.

### 13.8.1 General Access Profile

The General Access Profile (GAP) defines common modes and procedures used in all other profiles, thereby representing the minimum conformance requirements for Bluetooth devices. It ensures that all Bluetooth devices can establish contact with one another independent of the application profiles supported. The following functions make up the GAP:

- User interface;
- Modes;
- Security aspects;
- Idle mode procedures;
- Establishment procedures;
- Service discovery application.

### 13.8.2 Service Discovery Profile

The Service Discovery Profile (SDP) defines the protocols and procedures that will be used by a service discovery application on a device to locate services in other Bluetooth-certified devices using the Bluetooth SDP. With regard to this profile, the service discovery application is a specific, user-initiated application.

### 13.8.3 Transport Profiles

The profiles include specifications for serial port and legacy applications, as well as for generic OBEX applications and the exchange of any type of specified object.

### 13.8.4 Telephony Profiles

Two types of profiles are involved, one based on TCS and the other based on the serial port (AT Commands). Telephony control covers cordless telephony for residential systems and intercom applications such as walkie-talkie and local telephony. Serial port applications cover dial-up networking (cordless modem), faxing, and headsets.

### 13.8.5 Object Exchange Profiles

Object push profiles are used to push, pull, and exchange single objects, file transfer profiles effect the exchange of larger objects and files, and synchronization profiles enable synchronization of contacts and calendars between PCs and mobile electronic devices.

### 13.8.6 LAN Access

The LAN access profile with the point-to-point (PPP) protocol is used in PCs to access a LAN via a cordless network access point.

## 13.9 Competing Technologies

There is no single competitor covering the entire concept of the Bluetooth wireless technology, but in certain market segments other technologies exist. For cable replacement, the infrared standard IrDA has been around for some years and is quite well known and widespread. IrDA is faster than the Bluetooth wireless technology but is limited to point-to-point connections, and above all, it requires a clear line of sight. In the past, IrDA has had problems with incompatible standard implementations, a lesson that the Bluetooth SIG has learned.

Two other short-range radio technologies using frequency hopping technique reside in the 2.4-GHz band:

- Wireless LANs based on the IEEE 802.11 standard. The technology is used to replace a wired LAN throughout a building. The transmission capacity is high, and so is the number of simultaneous users. On the other hand, compared with Bluetooth, wireless LANs tend to be more expensive and power hungry. Because the hardware requires more space, wireless LANs are not suited for small mobile devices;

- HomeRF has many similarities with the Bluetooth wireless technology. HomeRF can operate ad hoc networks (data only) or be under the control of a connection point coordinating the system and providing a gateway to the telephone network (data and voice). The hop frequency is 8 hops/s, whereas a Bluetooth link hops at 1600 hops/s.

Ultra Wide Band radio is a new radio technology still under development. Short pulses are transmitted in a broad frequency range. The capacity is indicated to be high, whereas power consumption is expected to be low.

## 13.10 Conclusions

Bluetooth wireless technology is a worldwide de facto standard for small form factor, low-cost, short-range radio links between mobile PCs, mobile phones, and other portable devices. The primary function of Bluetooth is to connect computing and telecommunication devices without the need for cables. It delivers opportunities for rapid, ad hoc connections and the possibility of automatic connections between devices. Products are beginning to be delivered, and it is anticipated that by the year 2003 tens of millions of Bluetooth wireless technology devices will be around. A Cahners In-Stat study suggests that close to 700 million Bluetooth devices could be in operation by the year 2005. Bluetooth is available now and can be used for productive applications with immediate results. The devices and networks will continue to advance, but these future enhancements will improve the applications, not make them possible.

# Chapter 14

# Coexistence of IEEE 802.11b WLAN and Bluetooth WPAN

Stephen J. Shellhammer
*Symbol Technologies and IEEE 802.15.2 Chairman*

Both IEEE 802.11b wireless local area networks (WLANs) and Bluetooth wireless personal area networks (WPANs) operate in the same unlicensed frequency band. The use of a common frequency band for various incompatible wireless systems is fundamentally different than what has gone on in the past. Typically, as in a cellular phone system, the service provider has purchased the rights to use a specific frequency band in a geographic area. In a cellular system the band is often subdivided into two subbands so that two cellular providers can service the same market. However, the two providers do not share the exact same band.

The use of a common medium for multiple devices operating according to the same protocol has been around for years. For example, in Ethernet many computers are connected to a common cable. In systems that share the same medium one layer in the communication protocol is specifically designed to control the use of this shared medium. This layer is called the Medium Access Control (MAC) layer. But in all these cases the systems sharing the medium all conformed to the same specification. However, we now have totally different systems (i.e., 802.11b and Bluetooth) sharing the same medium.

As a result of these two wireless systems using the same shared medium there has been concern about the coexistence of these systems in a common location. The IEEE 802.15.2 WLAN/WPAN Coexistence Task Group was formed to address the issue of coexistence of these wireless systems. In this chapter, I describe the work of the IEEE task group, which includes modeling of the interference of IEEE 802.11b and Bluetooth and proposals for minimizing the mutual interference of these two systems.

## 14.1 Overview of WLAN/WPAN Coexistence in the 2.4-GHz ISM Band

IEEE 802.11b WLANs and Bluetooth WPANs both operate in the same unlicensed 2.4-GHz Industrial, Scientific, and Medical (ISM) frequency band. This frequency band has become very popular for deploying wireless networks because of its worldwide availability. Even though the band is unlicensed, it is regulated and there are rules that a device must follow to operate in this band. In addition to the WLAN and WPAN devices operating in this band, other devices such as cordless phones also use the band. Even the microwave oven in your home radiates in this band, and the RF leakage from the oven can interfere with WLAN and WPAN operation.

At the time the IEEE 802.15 Working Group on Wireless Personal Area Networks was formed, one of the concerns within the wireless community was whether the WLANs and WPANs that operate in the same frequency band would coexist peacefully. In other words, there was concern about the potential mutual interference between the WLAN and WPAN devices. To address that concern, the IEEE 802.15.2 WLAN/WPAN Coexistence Task Group was formed. When someone says that the two wireless networks coexist within the same band what he means is that the mutual interference between the two wireless networks does not prevent proper operation of each of the wireless networks. The IEEE 802.15.2 Task Group took on two tasks to address the issue of WLAN/WPAN coexistence. The first task was to develop a computer model of the mutual interference between 802.11b and Bluetooth, because these are the two most popular WLAN and WPAN technologies, respectively. This model is then used to predict the performance of the WLAN and WPAN networks in various user scenarios. Section 14.2 gives a brief overview of this model and shows examples of how the model can be used to predict network performance, in an interference environment.

The second task the group took on was to develop Coexistence Mechanisms, which are techniques to minimize the mutual interference between 802.11b and Bluetooth. There are two classes of coexistence mechanisms: collaborative and noncollaborative. A collaborative coexistence mechanism requires a communication link between the WLAN and WPAN networks so that the two networks can collaborate on a protocol to minimize interference between the two wireless networks. The most natural application of a collaborative coexistence mechanism is to ensure proper operation when both a WLAN and a WPAN are embedded inside the same physical unit, such as a laptop computer or a personal digital assistant (PDA). A noncollaborative coexistence mechanism is used when there is no communication link between the WLAN and WPAN networks. A good example of when a noncollaborative mechanism could be used is when a WLAN card is embedded in a laptop computer and a WPAN module is embedded into a PDA. In this example, because there is no link between the WLAN and WPAN networks the devices must adapt independently to reduce the mutual interference. The coexistence mechanisms being developed by the IEEE 802.15.2 coexistence task group are described in Section 14.2.

## 14.2 Mutual Interference of IEEE 802.11b WLAN and Bluetooth WPAN

The IEEE 802.15.2 WLAN/WPAN Coexistence Task Group developed a computer model of the mutual interference between an IEEE 802.11b WLAN and a Bluetooth WPAN. In this section I give a brief overview of the model and apply the model to several examples.

Computer networks are often described by dividing the network functionality into several layers. The most famous network model is the Open System Interconnection (OSI) model. IEEE 802 develops standards for the lower two layers of local and metropolitan area networks. Those lower two layers are referred to as the Physical (PHY) layer and the data link layer. The data link layer is divided into two sublayers: the Medium Access Control (MAC) layer and the logical link control layer. The design of the PHY and MAC layers has a major effect on the performance of the wireless network in the presence of interference. The IEEE 802.15.2 Interference Model includes PHY and MAC layer models for both the IEEE 802.11b WLAN and the Bluetooth WPAN. The terminology used in the Bluetooth specification is a little different than in the IEEE 802 networking standards, and hence the Bluetooth specification does not use the terms PHY or MAC. The RF layer of the Bluetooth specification is approximately the same as the physical layer. The Baseband and Link Manager layers of the Bluetooth specification are approximately the same as the MAC layer.

The PHY layer model of the Bluetooth WPAN is a baseband equivalent model of the Bluetooth RF layer. The model includes both a transmitter and receiver. The transmitter model consists of a Gaussian Frequency Shift Keying (GFSK) modulator operating at a data rate of 1 Mbit/s. The receiver model consists of a baseband filter and a frequency shift keying demodulator and detector. IEEE 802.11b supports four different data rates: 1, 2, 5.5, and 11 Mbit/s. At 1 Mbit/s the modulation is binary Differential Phase Shift Keying (DPSK). At 2 Mbit/s the symbol rate stays at 1 M symbol/s, however, there are 2 bits per symbol. The modulation used is Differential Quaternary Phase Shift Keying (DQPSK). At both data rates the signal is spread using a direct sequence spreading code at a rate of 11 Mchips/s. This gives the IEEE 802.11b signal a bandwidth of about 22 MHz. The higher data rates, 5.5 and 11 Mbit/s, use a more elaborate modulation scheme called Complementary Code Keying (CCK), while the bandwidth of the spectrum is maintained at about 22 MHz. The PHY layer model of IEEE 802.11b transmitter consists of a modulator using the appropriate modulation scheme at the various data rates. The PHY layer model of the IEEE 802.11b receiver consists of a baseband filter and the appropriate demodulator and detector at the various modulation schemes. The CCK demodulator and detector are quite complicated. For the details of the modeling, see Reference 1.

By using the PHY layer models of both Bluetooth and 802.11b we can calculate the Bit Error Rate (BER) of the two systems when they are both operating simultaneously. The BER of Bluetooth depends on the frequency offset between the Bluetooth channel and the middle of the IEEE 802.11b channel. That frequency offset is called $f_d$. Figure 14.1 shows the Bluetooth BER as a function of Signal-to-Interference Ratio (SIR) for various values of offset frequency. For a frequency offset of between 0 and 4 MHz the BER curve crosses the $10^{-3}$ level at SIR of about 5 dB. For an offset of more than 4 MHz the SIR for such a BER is less, because the power spectral density of the interfering 802.11b signal starts to drop off beyond 4 MHz. Because the power spectral density of 802.11b does not change significantly with data rate, the Bluetooth BER curve applies for all 802.11b data rates. In this simulation the Bluetooth radio is operating in interference limited environment with the Signal-to-Noise ratio (SNR) set at a high value of 35 dB.

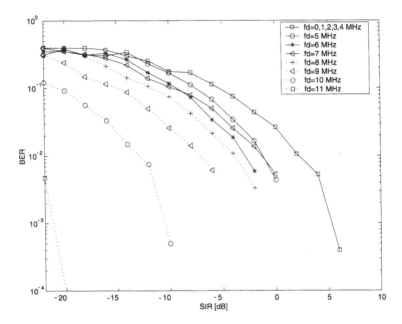

**Figure 14.1.** Bluetooth Bit Error Rate versus Signal-to-Interference Ratio in the presence of an IEEE 802.11b interferer.

Similarly, we can find the BER curve for IEEE 802.11b in the presence of a Bluetooth interferer. Because IEEE 802.11b has four data rates we need to find the BER curve for the four different data rates. Figure 14.2 shows the BER curves for IEEE 802.11b operating at 1 Mbit/s versus SIR for various frequency offsets between the middle of the 802.11b channel and the Bluetooth channel. The worst case frequency offset is 1 MHz, not 0 MHz as one might first expect. The reason for this is the direct sequence spreading code that is used has a null in the middle of its spectrum, which suppresses interference. The SIR is quite low because the 802.11b operating at 1Mbit/s benefits from the 10.4-dB direct-sequence processing gain. As the frequency offset increases SIR level for a fixed BER decreases, just as in the Bluetooth case. The noise level is quite low, with the SNR set at 35 dB.

We can also simulate the 802.11b receiver operating at 11 Mbit/s using CCK modulation. In this case, because the rate of the spread spectrum chipping sequence is equal to the data rate, the system no longer benefits from the 10.4-dB processing gain that it has at 1 Mbit/s. Of course, the benefit the user has is the higher data rate for his application. Figure 14.3 shows the 802.11b BER curve versus SIR when operating at 11 Mbit/s. Once again the SNR is set at 35 dB. Note that the shape of the curves are similar to those of the 1 Mbit/s case; however, at 11 Mbit/s the SIR needs to be about 6 dB higher, for the same BER. When operating at 11 Mbit/s 802.11b does not benefit from the 10.4-dB processing gain; however, it has an approximately 4-dB advantage because of the more sophisticated CCK modulation. Overall, the 11 Mbit/s signal needs a 6-dB-higher SIR compared with the 1 Mbit/s signal to maintain the same BER.

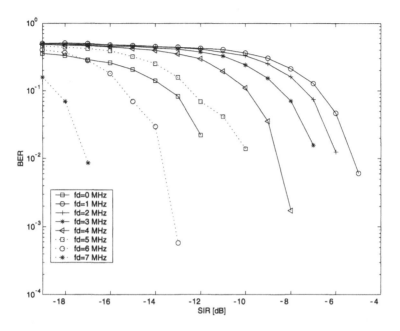

**Figure 14.2.** IEEE 802.11b Bit Error Rate operating at 1 Mbit/s versus Signal-to-Interference Ratio in the presence of a Bluetooth interferer.

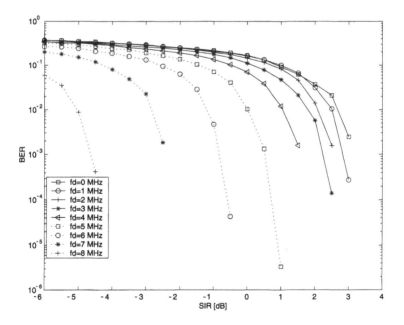

**Figure 14.3.** IEEE 802.11b Bit Error Rate operating at 11 Mbit/s versus Signal-to-Interference Ratio in the presence of a Bluetooth interferer.

To find the BER for various application scenarios we need the Signal-to-Interference Ratio (SIR) at the various receivers. The SIR is based on several parameters: the transmit power of both the WLAN and WPAN nodes, distances from these transmitters to the receiver of interest, and finally the path loss of the RF signal as it propagates from the transmitters to the receiver under evaluation.

The formula for the path loss that is used in this simulation is a two-slope linear path loss model. The path loss formula has a path loss coefficient of 20 dB up to 8 m and then changes to a path loss coefficient of 33 dB starting at 8 m. In free space the path loss coefficient is 20 dB. However, inside a building with cubicle walls or other furniture that can block the line of sight RF path, the path loss coefficient is higher. Thus this model assumes line-of-sight between nodes out to about 8 m and then some blockage or clutter starting at 8 m. The formula for the path loss that is used in the model is given by,

$$L = 40.2 + 20\log(d) \qquad d < 8 \text{ m}$$
$$L = 58.5 + 33\log(d/8) \qquad d > 8 \text{ m}$$

where the path loss, $L$, is measured in dB and the distance, $d$, is measured in meters.

Given a geometric distribution of WLAN and WPAN nodes the SIR is calculated at the various receivers. From the BER curves the BER of the WLAN and WPAN nodes can be calculated for intervals in which the nodes are simultaneously operating. For example, if the 802.11b Access Point is sending data to an 802.11b Mobile Unit while a nearby Bluetooth node is transmitting we can calculate the BER at the 802.11b Mobile Unit. This information is used by the MAC layer simulation to determine whether the 802.11b packet is in error and will need to be retransmitted, resulting in a drop in WLAN throughput. Next we will introduce the MAC layer simulation and give some simulation results.

The MAC layer simulation was developed using a network-modeling tool called OPNET[TM]. The IEEE MAC layer model was available in OPNET, so the IEEE 802.15.2 task group only needed to develop a model for the Bluetooth MAC (i.e., baseband and link manager). The Bluetooth MAC layer simulation includes the various Bluetooth packet types for both data and voice traffic. The MAC layer model calculates when the 802.11b and Bluetooth packets overlap in time and what the frequency offset between the two wireless networks is for that packet overlap. The PHY layer model calculates the BER for both the WLAN and WPAN networks. Given the BER for the time interval of overlapping transmission the MAC layer simulation calculates various performance metrics. These performance metrics include the Packet Error Rate (PER), the network throughput, and the network latency at the top of the MAC layer.

To illustrate the utility of the complete IEEE 802.15.2 mutual interference model let us apply it to several application scenarios. First, let us look at the performance of the Bluetooth voice link in the presence of a nearby WLAN mobile unit. Because the Bluetooth voice link does not include any packet retransmission the primary performance metric of interest is the PER. As the PER increases, the voice quality degrades. In Reference 2, Golmie applies the IEEE simulation to the case of two Bluetooth nodes separated by 1 m exchanging voice traffic. Both Bluetooth HV1 and HV3 data links are considered. The resulting PER (or loss rate) is plotted in Figure 14.4. The independent variable is the distance between one of the Bluetooth nodes and an interfering WLAN mobile unit. As you can see the PER approaches 15% as the 802.11b mobile unit comes near the Bluetooth node. A PER of that magnitude will significantly degrade the voice quality.

The robustness of the Bluetooth voice link can be significantly improved by adding packet level redundancy, instead of bit level redundancy. The reason for this is that in an interference environment the bit errors within a packet are not independent; however, since each packet is sent on a different frequency, packet errors are independent. A proposal was made within the IEEE 802.15.2 task group for a new Bluetooth voice link entitled: Synchronous Connection-Oriented with Repeated Transmission (SCORT) Link [3]. In this proposal a simple packet redundancy was proposed, fitting with the Bluetooth goal of simplicity and low complexity. The repeated packet transmission can easily be converted into an Automatic Repeat Request (ARQ) technique by adding an acknowledgment bit in each voice packet. Figure 14.5 shows the significant reduction in PER with packet redundancy. Note how the packet loss rate drops from around 15% to around 3% with SCORT.

If the Bluetooth link under consideration were a data link instead of a voice link, then a 15% packet error rate would correspond to throughput of about 85% of peak Bluetooth throughput, which is not a dramatic reduction in throughput. So, as you can see, the Bluetooth voice link is more susceptible to interference than a Bluetooth data link. Similarly, we can investigate the performance of an IEEE 802.11b WLAN in the presence of a Bluetooth interferer [4]. Figure 14.6 shows the packet loss of 802.11b WLAN at both 1 Mbit/s and 11 Mbit/s, in the presence of a Bluetooth interferer. The Bluetooth WPAN is sending HV1 voice traffic. The distance between the 802.11b access point and the 802.11b mobile unit is fixed at 15 m, which is much less than the typical range of an 802.11b WLAN.

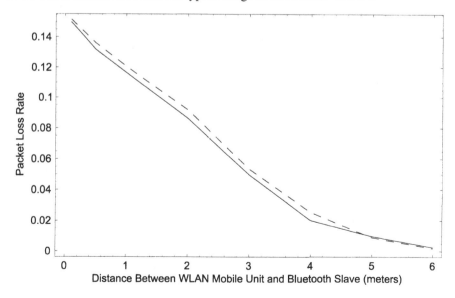

**Figure 14.4.** Bluetooth packet loss rate versus separation of Bluetooth and WLAN node. Solid line: HV1, dashed line: HV3.

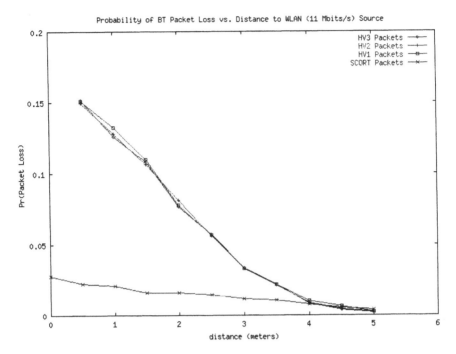

**Figure 14.5.** Bluetooth packet loss rate for standard Bluetooth voice links and proposed SCORT link.

At 1 Mbit/s the 802.11b packet loss rate can reach 65%, which corresponds to a throughput of about 35% of the standard 802.11b throughput, operating at 1 Mbit/s. As the SIR increases the BER drops off sharply. The cause of this sharp drop in BER is the processing gain of the direct sequence spreading code. At 11 Mbit/s the maximum packet loss rate is only around 30%. The reason for this is that the duration of the packets is shorter with a higher data rate. The probability of collision drops as the time duration of the packet decreases; therefore, the worst case packet loss is worse at 1 Mbit/s than at 11 Mbit/s. However, the 11 Mbit/s operation does not benefit from the high spread spectrum processing gain of the 1 Mbit/s operation.

## 14.3 Coexistence Mechanisms for Reducing WLAN/WPAN Mutual Interference

Now that we have seen how IEEE 802.11b and Bluetooth perform in an interference environment, let us look at methods of reducing the mutual interference of these two wireless networks. IEEE 802.15.2 will be publishing a Recommended Practice of these coexistence mechanisms. As explained in Section 14.1, there are two classes of coexistence mechanisms: collaborative and noncollaborative. A collaborative coexistence mechanism is used when 802.11b and Bluetooth are to be embedded into the same physical unit and have a physical communication link between them.

You can think of this as a method of ensuring proper operation of a dual-mode 802.11b/Bluetooth radio. A noncollaborative mechanism is to be used when 802.11b and Bluetooth are in separate physical units and hence there is no communication link between the two wireless networks.

Both the collaborative coexistence mechanisms consist of a packet scheduling technique to ensure that the 802.11b and Bluetooth packets do not collide. The first technique, called Alternating Wireless Medium Access (AWMA), takes the time interval between two adjacent 802.11b beacons and subdivides it into two subintervals: one for 801.11b and one for Bluetooth [5]. Figure 14.7 shows the two intervals, one for the WLAN and one for the WPAN. The WLAN interval begins just before the WLAN Target Beacon Transmit Time (TBTT). The timing is controlled by the 802.11b MAC, which sends a *Medium Free* signal to the Bluetooth unit to signal to the Bluetooth unit when it can use the medium (i.e., air waves), as shown in Figure 14.8. One of the benefits of AWMA is that all the 802.11b mobile units attached to the same access point share a common clock and hence have the exact same WLAN and WPAN intervals. Not only is there no interference within the mobile unit containing both 802.11b and Bluetooth, there is also no interference from other nearby mobile units. In other words, the Bluetooth in one laptop computer will not transmit while the 802.11b WLAN in the laptop next to it is transmitting or receiving. This ensures that there is no 802.11b and Bluetooth interference between the two laptops. This coexistence mechanism can be thought of as a coordinated MAC layer for both radios.

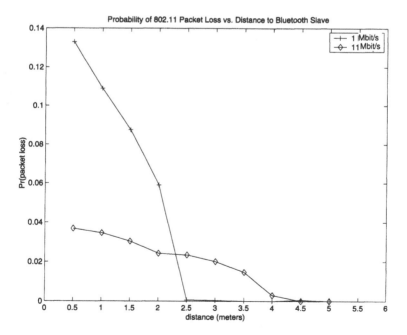

**Figure 14.6.** Packet loss of an 802.11b WLAN at both 1 Mbit/s and 11 Mbit/s, in the presence of a Bluetooth interferer.

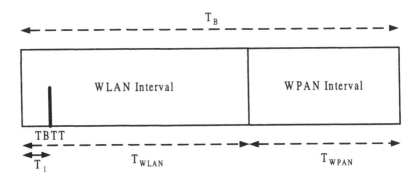

**Figure 14.7.** Timing of the WLAN and WPAN subintervals in AWMA.

The second collaborative coexistence technique, called MEHTA [5], acts like a wireless traffic cop, deciding which wireless system can transmit on a packet-by-packet basis. Each wireless system, 802.11b and Bluetooth, submits requests for access to the medium. The MEHTA packet scheduler sends a message to each wireless system when it is allowed to transmit. If one system requests the medium and the other does not, then that system is allowed access to the medium. However, if both systems request access such that there would be a collision, then MEHTA will select which system is given access based on a priority scheme. For example, if 802.11b is sending data traffic and Bluetooth is sending voice traffic, then MEHTA will give priority to Bluetooth in any request that would lead to a collision. In this case 802.11b must wait until the Bluetooth pack has completed before it can send its data packet. If 802.11b and Bluetooth are both sending data traffic, then the priority scheme can be set up in several ways. Either one system, like 802.11b, can always have priority, because its data may be more time critical, or a probability can be assigned to each system, and that system is selected with that probability when a collision would of occurred. For example, if 802.11b and Bluetooth are each assigned a probability of 0.5, then every time a collision would have occurred MEHTA generates a binary random number (i.e., it flips a coin). If the random number is 1, MEHTA gives access to 802.11b; if the number is 0, then MEHTA gives access to Bluetooth. A block diagram of the MEHTA system is shown in Figure 14.9.

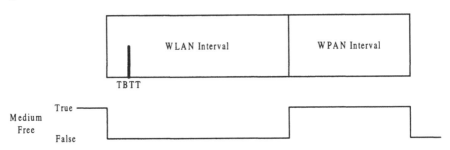

**Figure 14.8.** Timing of Medium Free signal in AWMA.

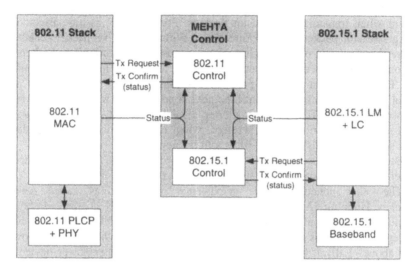

**Figure 14.9.** Structure of MEHTA control.

There are two types of noncollaborative coexistence mechanisms: adaptive packet selection and scheduling and adaptive frequency hopping. Both techniques are intended to be used when 802.11b and Bluetooth are operating is separate physical units.

Adaptive packet selection and scheduling [6] is a technique within the Bluetooth WPAN to decide, based on knowledge of the interference environment, which packet type to use and also when to transmit and when to wait. There is assumed to be a frequency-static interferer, like 802.11b, which is corrupting some of the 79 Bluetooth channels. Typically, an 802.11b radio uses 22 MHz, which covers about 23 Bluetooth channels. The Bluetooth WPAN is augmented with a subsystem for classifying channels as *good* (not occupied by the frequency static interferer) and *bad* (occupied by the frequency static interferer). This channel classification can be based on a number of channel measurements including Bluetooth PER and background Receive Signal Strength Indicator (RSSI). This channel classification can be done only within the Bluetooth master, to save costs, or can be done in both the master and the slaves. If it is done in both master and slaves, then the slaves need to communicate their channel classifications to the master.

Once the Bluetooth master knows which channels are good and which are bad it is possible for the Bluetooth master to schedule packet transmissions so as to avoid the bad channels. As shown in Figure 14.10, the PER can be eliminated by scheduling packet transmission only on good channels. Because Bluetooth is not transmitting on bad channels, packet collisions with 802.11b have been eliminated and hence, 802.11b throughput is maintained. Clearly, the effectiveness of this technique depends on the accuracy of classification of the channels as either good or bad. Because Bluetooth delays transmission until the next good channel, its throughput is reduced and latency is affected. When sending single-slot packets, if the channel is bad, the master will delay transmission until the next good channel, which will typically lead to a two-slot delay. Thus, there is a slight increase in overall latency when single-slot packets are transmitted. However, the story is different when transmitting multislot packets. For example, assume Bluetooth is sending five-slot packets in both directions. Without adaptive packet scheduling, suppose the master is on a bad channel but sends a five-slot packet. If there is a collision, that packet will not be retransmitted for ten slots. This leads to significant delay. However, when adaptive packet scheduling is used, the master on a bad channel will delay transmission until the next good channel, which typically results in a two-slot delay. This delay is shorter than the ten-slot delay when adaptive packet scheduling is not used. Therefore, when using multislot packets, adaptive packet scheduling tends to reduce Bluetooth latency.

Bluetooth can also schedule the types of packets it uses so that when the channel is good more data can be sent and will get through more often.

The other noncollaborative coexistence mechanism is called Adaptive Frequency Hopping (AFH). In AFH the Bluetooth WPAN changes its hopping sequence to "hop around" frequency-static interferers like 802.11b. Under current FCC regulations this is only allowed for Bluetooth devices whose transmit power is below about –1 dBm. Under the current FCC rules for higher-power devices you must use at lease 75 channels, which are regulated under FCC Part 15.247. However, in May 2001 the FCC issued a Notice of Proposed Rulemaking (NPRM) [7], which if approved would reduce the number of used channels to 15, "for any system that uses adaptive hopping techniques." The RF transmit power is limited to 125 mW, which is higher than any current Bluetooth system.

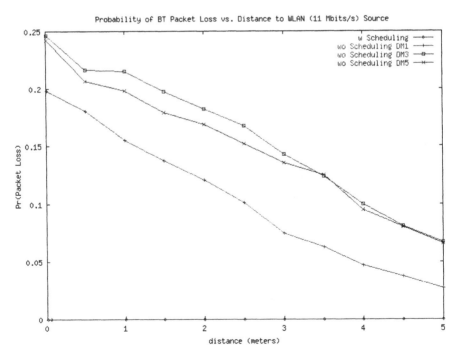

**Figure 14.10.** Packet loss rate in Bluetooth with and without adaptive packet scheduling.

In AFH there is a channel classification procedure just as in adaptive packet selection and scheduling. Once the channels have been classified the hopping sequence is modified to avoid the bad channels [8]. A device called the frequency remapper performs the function of modifying the hopping sequence. The original hopping frequency generator is left unmodified. The output of the original hopping generator is fed into the frequency remapper, which either passes the selected frequency through unmodified if it is a good channel or remaps a bad channel onto a good channel. A Bluetooth master with AFH capability will still be able to communicate to a legacy Bluetooth slave on good channels that have not been remapped.

There may be a situation in which there are fewer good channels than the minimum number of allowed channels set by local regulations. For example, if the FCC NPRM is approved, then the minimum number of used channels in the United States would be set at 15 channels. In Europe the minimum number of used channels is set at 20. If for example, the Bluetooth WPAN is operating in a room with three 802.11b access points all of which are operating on a different channel, then the number of good channels would be about $79 - (3 \times 23) = 10$. In this case there are fewer available good channels than the number of channels that must be used by regulation, both in the US and in Europe. Hence, some bad channel must be used.

If any remaining bad channels are to be used, then the good and bad channels are regrouped such that the good channels are clustered together and the bad channels are also clustered together. The partition sequence generator is the function that groups the good and bad channels in clusters. Figure 14.11 shows the block diagram of AFH frequency generation circuit. Bluetooth voice traffic can be scheduled to operate over the available good channels, giving it priority over the data traffic.

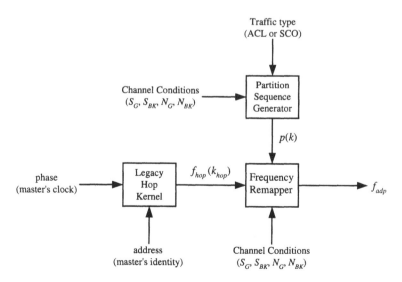

**Figure 14.11.** Adaptive Frequency Hopping frequency generation circuit.

AFH not only prevents packet collisions with frequency static interferers like 802.11b, but because Bluetooth does not defer transmission, the Bluetooth throughput is maintained near its maximum. The IEEE 802.15.2 WLAN/WPAN Coexistence Task Group is currently working with the Bluetooth Special Interest Group (SIG) to agree on a common AFH specification.

## 14.4 Conclusions

In this chapter, I have summarized the work of the IEEE 802.15.2 WLAN/WPAN Coexistence Task Group. From the mutual interference model we can make several observations. As one might expect, if the two wireless systems are placed nearby one another the mutual interference can impact the network performance. Obviously, the closer the devices are to one another the more significant the interference. The Bluetooth voice link is more susceptible to interference than the Bluetooth data link primarily because the voice link does not support any form of packet retransmission. The IEEE 802.11b throughput can drop significantly if a Bluetooth device is too close. Dropping the IEEE 802.11b data rate does not always help, because at the lower data rates the packets are longer in duration and hence the packet error rate can even increase. However, beyond a certain distance the processing gain at 1 Mbit/s does significantly reduce the packet error rate. Several coexistence mechanisms have been described, which will be included in the IEEE 802.15.2 Recommended Practice, that significantly improve the performance of both 802.11b and Bluetooth when operating in a common area.

## References

[1]    J. Lansford, A. P. Stephens, R. E. Van Dyck, and A. Soltanian, *Combined Text for Clause 6*, IEEE 802.15-01/444r1, October 2001.

[2]    N. Golmie, *Using a Combined MAC and PHY Simulation Model to Measure WLAN Interference on Bluetooth (Part I)*, IEEE 802.15-00/388r0, November 2000.

[3]    S. J. Shellhammer, *SCORT—An Alternative to the Bluetooth SCO Link for Voice Operation in an Interference Environment*, IEEE 802.15-01/145r1, March 2001.

[4]    N. Golmie, R. E. Van Dyck, and A. Soltanian, *Performance Evaluation of Bluetooth and IEEE 802.11 Devices Operating in the 2.4 GHz ISM Band*, Proceedings of the Fourth ACM International Workshop on Modeling, Analysis, and Simulation of Wireless and Mobile Systems, MSWIM'01, July 2001, Rome, Italy.

[5] J. Lansford, A. P. Stephens, R. E. Van Dyck, and S. Shellhammer, *Combined Text for Collaborative Coexistence Mechanism Clause,* IEEE 802.15-01/506r0, November 2001.

[6] N. Golmie and J. Liang, *Non-Collaborative MAC Mechanisms,* IEEE 802.15-01/316r0, July 2001.

[7] FCC Notice of Proposed Rulemaking, *Amendment of Part 15 of the Commission's Rules Regarding Spread Spectrum Devices,* Docket Number 01-158, May 11, 2001.

[8] A. Batra, H. K. Chen, and Hongbing Gan, *Adaptive Frequency Hopping,* IEEE 802.15-01/491r0, November 2001.

# Chapter 15

# An Introduction to Ultra Wide Band Wireless Technology

Kazimierz Siwiak and Laura L. Huckabee
*Time Domain Corporation*[1]

Ultra Wide Band (UWB) signaling is a modern wireless technique whereby very short baseband signals are transmitted and received without a radio frequency (RF) carrier. The technique reuses previously allocated RF bands by spreading the impulse energy thinly in a wide spectrum, thus hiding signals below the noise floor of conventional receivers. There are several methods of generating and radiating UWB signals, including "low-duty cycle UWB" implemented as Time-Modulated UWB™ (TM-UWB™), and high-duty cycle "direct sequence phase coded UWB" (DS-UWB). Unique ways of coding and positioning the impulses, as in Transmitted-Reference Delay hopped UWB (TRD-UWB) can result in the efficient raking of the indoor multipath energy. In this chapter, the technology basics for time position coded impulse UWB, the direct sequence coded UWB, and the transmitted reference UWB approaches are explained. Applications of UWB devices are presented, and potential uses of UWB are described. It is shown that short-pulse low-power techniques have enabled practical through-wall radars, centimeter-precision 3-D positioning, and communications capabilities at the high data rates and with exceptional spatial capacities.

## 15.1 Introduction

Ultra Wide Band (UWB) signaling is essentially the art of generating, modulating, emitting, and detecting baseband digital signals, which inherently occupy large bandwidths. Impulse transmissions can be said to date back to the infancy of wireless technology. They include the experiments of Heinrich Hertz in the 1880s and later the 100-year-old spark gap "impulse" transmissions of Guglielmo Marconi, who in 1901 sent the first-ever over-the-horizon wireless transmission from the Isle of Wight to Cornwall on the British mainland. Early radio comprised only passive electrical components and no tubes or transistors and hence lacked the

---

[1] Time-Modulated UWB™ and TM-UWB™ are trademarks of Time Domain Corporation, Huntsville, AL.

means to deal efficiently with short transient impulses, so radio subsequently developed along narrow-band frequency-selective analog techniques. This led to voice broadcasting and telephony—and recently transitioned to digital telephony and data. Through the years, a small cadre of scientists have worked to develop and refine impulse technologies. Before 1970 the primary focus was on impulse radar techniques and on government-sponsored projects. In late 1970s and early 1980s, however, digital techniques began to mature to the point where the practicality of modern low-power-impulse radio communications can be demonstrated with the impulse time coding and modulation approach. Digital impulse radio, [see 1–5], the modern echo of Marconi's century-old transmissions, now emerges as UWB radio. Alternate methods of generating signals having UWB characteristics are being developed, including the use of continuous streams of pseudonoise (PN) coded impulses that resemble CDMA signaling at a chip rate commensurate with the emission center frequency. The industry is now moving to commercial deployment.

UWB signaling is more nearly characterized by transient circuit responses, whereas conventional radio tends to deal with the steady state. We will describe impulse transmitting and receiving systems. Impulse propagation, especially indoors, also differs significantly from that of narrow-band carrier-based systems in that multipath is described by sometimes overlapping but otherwise distinct short impulses rather than continuous sine waves forming complex interference patterns. Finally, we will illustrate many applications of UWB systems.

## 15.2 Coded UWB Impulses and Impulse Streams

UWB radio is the transmission and reception of ultra-short electromagnetic energy impulses and is the generic term describing radio systems having very large instantaneous bandwidths. The U.S. Federal Communications Commission (FCC), for example, has tentatively defined UWB systems as "having bandwidths greater than 25% of the center frequency measure at the 10 dB down points" or "RF bandwidths greater than 1.5 GHz," whichever is smaller. There are several methods of generating, radiating, and receiving such UWB signals, including TM-UWB, DS-UWB, and TRD-UWB. Wide spectra are generated in all methods; however, the radio techniques, signal characteristics, and application capabilities vary considerably.

Developers have perfected various ways for creating and receiving these signals, and for encoding information in the transmissions. Pulses can be sent individually, in bursts, or in continuous streams and can encode information in pulse amplitude, phase, and pulse position. Modulations vary from simple pulse position to a more energy efficient pulse polarity [see 6] and to the very energy efficient *M-ary* (multilevel) pulse position modulation. Modern UWB radio is characterized by very low-power transmission- and, by virtue of those wide bandwidths (greater than a gigahertz), extremely low-power spectral densities. The emissions are targeted to be below −41.25 dBm/MHz Effective Isotropically Radiated Power (EIRP) in bands below 960 MHz, between 1.99 and 10.6 GHz, and near 24 GHz under US CFR-47 Part 15 as defined in a new Report and Order issued in 2002.

Three commercially useful UWB communications techniques exemplify the wide range of implementation possibilities: TM-UWB, DS-UWB, and TRD-UWB.

All systems use transient switching techniques to generate brief (typically subnanosecond) impulses or "monocycles" having a small number of zero crossings. The impulses are radiated by specialized wide-band antennas [see, for example, 7] that may be designed in new and different ways [8].

TM-UWB impulses are transmitted at high rates, in the millions to tens of millions of impulses per second. However, the pulses are not necessarily evenly spaced in time, but rather they may be spaced at random or pseudorandom time intervals. The process creates a noiselike signal in both the time and frequency domains. Data modulation is applied by further dithering the timing of the pulse transmissions, by signal polarity, and perhaps pulse amplitude. A coherent correlation-type receiver and integrator converts the UWB pulses to a baseband digital signal that has a bandwidth commensurate with the data rate. The correlation operation and subsequent integration filtering provide significant processing gain that is effective against interference and jamming. Time coding of the pulses allows for channelization, whereas the time dithering, pulse position, and signal polarity provide the modulation. UWB systems built around this technique, and operating at very low RF power levels, have demonstrated very impressive short- and long-range data links, positioning measurements accurate to within a few centimeters, and high-performance through-wall motion-sensing radars.

DS-UWB uses high-duty cycle phase-coded sequences of wide band impulses transmitted at gigahertz rates. Sequences of tens to thousands of impulse "chips" encode data bits in scalable data rates from a one to hundreds of Mbit/s. The modulation is by pulse polarity and resembles a base band CDMA system with the chipping rate commensurate with the center frequency. The PN encoding per data bit provides a measure of multipath delay spread tolerance, allows for channelization, and provides processing gain against interferers. A direct sequence-type of receiver can be used to correlate with the PN code and convert the integrated impulses to data rate bandwidths.

TRD-UWB employs impulse pairs that are differentially polarity encoded by the data with the pulse pairs transmitted having a precise spacing $D$. The receiver comprises a correlator with one input fed directly and another input delayed by $D$, similar to the conventional DPSK system, except rather than integrating over a bit time, here the integration time is commensurate with multipath decay time. The differentially encoded delayed reference impulse and its data impulse are affected the same way by multipath; hence, the delayed-reference detection and integration operation behave like a near-perfect rake receiver capturing a large percentage of the multipath induced signal replicas.

## 15.3 UWB Technology Basics

UWB spectra can be generated in several different ways, for example, by TM-UWB, which uses low-duty cycle impulses, by DS-UWB, which uses high-duty cycle waveforms that are direct sequence phase modulated, and by coded pulse pairs in TRD-UWB. Ultra-short impulse waveforms are common to both technologies.

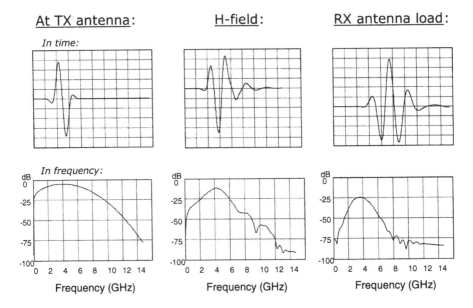

**Figure 15.1.** Source, emitted and received monopulses. After [9].

The monocycle waveform applied to the transmitting antenna and represented in Figure 15.1 along with its frequency spectrum, is the most basic element of UWB signaling. The monocycle center frequency and the bandwidth are completely dependent upon the monocycle's width. Actual radiated waveforms and spectra as well as the received waveform and spectra, like the *H*-field shown in Figure 15.1, are further shaped by the band pass and transient response characteristics of the transmitting antenna. The waveform and its spectrum change yet again in the receiving antenna load, reflecting the transient impulse response of the entire UWB link.

If the pulses had been sent at a regular interval without PN encoding, the resulting spectrum will contain "comb lines" separated by the pulse repetition rate. The resulting peak power in the comb lines will undesirably limit the total transmit power as measured in any 1-MHz bandwidth. Instead, to make the spectrum more noiselike and provide for channelization, the monocycle impulses are pseudorandomly placed within each time frame.

TM-UWB employs PN encoded time dithering to place pulses to picosecond accuracy within a time window equal to the inverse of the average pulse repetition rate. Figure 15.2 illustrates a "pulse train" that has been PN time coded and shows the resulting frequency spectrum. The PN coding comprises a pseudorandom time shift within each frame time. DS-UWB, on the other hand, uses PN codes to polarity-modulate pulse sequences that are closely spaced and at regular intervals. TRD-UWB uses precisely spaced pulse pairs that are polarity modulated. The resulting spectra of DS-UWB and TRD-UWB are similar to those of TM-UWB.

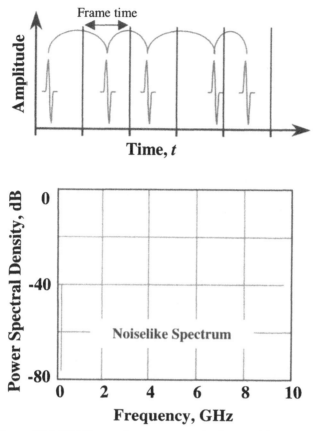

**Figure 15.2.** UWB waveform sequence in time and frequency.

### 15.3.1 TM-UWB Technology

TM-UWB transmitters emit ultra-short monocycle waveforms with tightly controlled pulse-to-pulse intervals. The waveform pulse widths are typically between 0.2 and 1 ns, corresponding to center frequencies between 5 and 1 GHz, with pulse-to-pulse intervals of between 25 and 1,000 ns. The systems typically use pulse position and polarity modulation. The pulse-to-pulse interval is varied on a pulse-by-pulse basis in accordance with two components: an information signal and a channel code. The TM-UWB receiver directly converts the received RF signal into a baseband digital or analog output signal. A front-end correlator coherently converts the electromagnetic pulse train to a baseband signal in one stage. There is no intermediate frequency stage, greatly reducing complexity. A single bit of information may spread over multiple monocycles, providing a way of scaling the energy content of a data bit with the data rate. The receiver coherently sums the proper number of pulses to recover the transmitted information.

TM-UWB systems use pulse position modulation by positioning the pulse one quarter cycle (60 ps for a 240-ps pulse) early or late relative to the nominal PN-coded location or by pulse polarity. Furthermore, multilevel pulse position modulation may be used to provide enhanced bit energy-to-noise ratio performance. Modulation further "smoothes" the signal spectrum, thus making the signals less detectable.

### 15.3.2 A TM-UWB Transmitter

Figure 15.3 shows a high-level block diagram of a TM-UWB transmitter. The transmitter has no power amplifier, but rather, pulses are generated at the required power. A precision programmable delay implements the PN time coding and time modulation. Alternatively or in addition, modulation can be encoded in pulse polarity. The precise timing capability of the timer operation (several-picoseconds resolution) enables not only precise time modulation and precise PN encoding but also precision distance determination. The picosecond-precision timer, implemented in an integrated circuit, is a key technological component of the TM-UWB system.

### 15.3.3 A TM-UWB Receiver

The receiver shown in Figure 15.4 resembles the transmitter, except that the pulse generator feeds the multiplier within the correlator. The performance of this type of correlator receiver is described in Reference 9. Baseband signal processing extracts the modulation and controls signal acquisition and tracking. Baseband signal processing also drives a tracking loop that locks onto the time coded sequence. Modulation is decoded as either an "early" or a "late" pulse in time modulation and/or as a positive or negative pulse in polarity modulation. Different PN time codes are used for channelization. Precise pulse timing inherently enables exceptional positioning and location capabilities in TM-UWB communications systems.

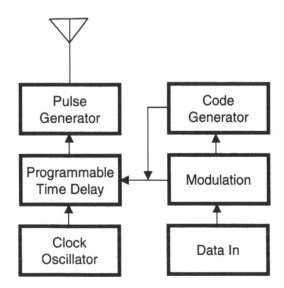

**Figure 15.3.** A TM-UWB transmitter.

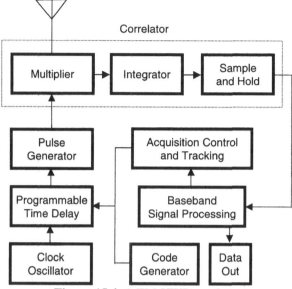

**Figure 15.4.** A TM-UWB receiver.

### 15.3.4 DS-UWB Technology

A second method of generating useful signals having UWB spectra comprises a DS-UWB approach not unlike an RF carrier-based CDMA system. Impulse sequences at duty cycles approaching that of a sine wave carrier are direct-sequence polarity (like binary phase shift keying) modulated. The PN sequence provides smoothing, channelization, and modulation. The chipping rate is some fraction $1/N$ of the "carrier" center frequency. For illustration, Figure 15.5 shows the approximate spectral envelope of a 4-GHz impulse sequence that is DS modulated by a zero mean PN code for the cases $N = 1$ and $N = 2$. Actual PN sequences are relatively short, and the spectra contain more features, as depicted in Figure 15.2. Both signals in Figure 15.5 have the same power in a 1-MHz bandwidth at 4 GHz, but the $N = 1$ signal carries the greater total power in the spectrum. The total power and occupied bandwidth can be traded off subject to regulatory emissions limits.

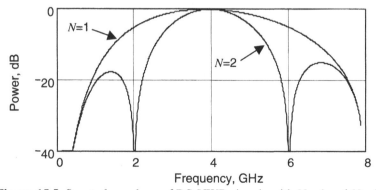

**Figure 15.5.** Spectral envelope of DS-UWB signals with $N = 1$ and $N = 2$.

### 15.3.5 TRD-UWB Technology

A method of transmitting and receiving impulses that implements a near-perfect rake receiver is exemplified by TRD-UWB and described in Reference 10. The method employs differentially encoded impulse pairs dent at a precise spacing $D$. The system is shown in the simplified block diagram of Figure 15.6. The transmitter sends a pair of pulses separated by a delay $D$ and differentially encoded by pulse polarity. The pulses, including propagation-induced multipath replicas, are received and detected with a correlator with one input fed directly and another input delayed by $D$. The receiver resembles a conventional DPSK receiver. Integration is over a time sufficiently long to rake in a significant amount of the multipath energy. The TRD-UWB receiver tends to behave like a near-perfect rake receiver capturing a large percentage of the multipath-induced signal replicas.

## 15.4 UWB Signal Propagation

Conventional narrow-band radio is plagued by multipath within and around buildings, has difficulty with precision tracking in locations that have significant multipath, and has difficulty resolving targets in environments with lots of clutter. Furthermore, multipath and diffraction phenomena conspire to degrade the propagation characteristics of "continuous wave" conventional radio, especially inside buildings. Some UWB techniques, on the other hand, thrive indoors, enable positioning accuracies to better than a few centimeters, and generally follow a free space propagation law [11]. Further indoor channel characteristics are described in References 12–14. Additionally, UWB is potentially more difficult to detect than traditional radio. UWB systems, for example, transmit millions to billions of coded pulses per second at emission levels below the noise floor of conventional narrow-band receivers. These transmissions have a very low extant RF signature, providing intrinsically secure transmissions with low probability of detection and low probability of interception.

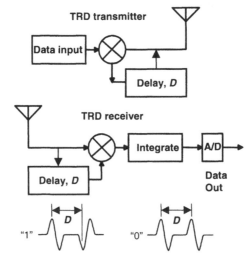

**Figure 15.6.** A TRD-UWB transmitter and receiver.

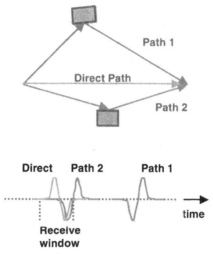

**Figure 15.7.** Direct and reflected impulse paths.

Multipath fading, the bane of RF communications, is the result of coherent interaction of sinusoidal signals arriving by many paths. Spread spectrum IS-95 cellular and PCS systems with a 1.228-MHz spreading bandwidth can resolve multipath signals having differential delays of slightly less than 1 μs. Some communications channels, particularly outdoors, can have rms delay spreads measuring many microseconds, so some multipath components can be resolved and received with rake techniques. However, in-building communications channels exhibit multipath differential delays and rms delay spreads in the several tens of nanoseconds and cannot be resolved in the relatively narrow IS-95 channel. Systems like IS-95 must therefore contend with significant Rayleigh fading, which requires signals up to tens of decibels above the static signal level for a given measure of performance.

Properly designed UWB systems can have bandwidths exceeding a gigahertz and are capable of resolving multipath components with differential delays of less than a nanosecond. Figure 15.7 shows a multipath scenario for a direct impulse path and reflected paths 1 and 2. When reflected path 1 is more than a pulse length longer than the direct path, the reflected impulse appears as a distinct delayed signal carrying additional energy to the receiver. When the differential delay between reflected path 2 and the direct path is smaller than the pulse length, the replica pulse overlaps the direct pulse. As long as the overlap is less than half of the pulse length, this delayed pulse contributes positively to received energy, because the reflected impulse is polarity inverted by the reflection. No margin for Rayleigh fading is needed because there are no overlapping sine waves to introduce destructive interference.

Multiple delayed replica impulses can be integrated or rake-received to provide *gain* over a single direct path in the multipath environment. Raking in all of the multipath-delayed signal replicas will result, on the average, in free space-like propagation indoors, potentially providing a significant improvement in performance over an equal power sine wave system. Wide signal bandwidths also mean that current narrow-band users of the spectrum "see" only a slight, if any, increase in the background noise—and then only if within a few meters of an active UWB source [see, for example, 14]. Wide signals also mean large available processing gains for the UWB system to work in the presence of high-power narrow band users.

## 15.5 The Future of UWB

UWB technology uniquely harnesses an ultra-wide band of spectrum to provide high-bandwidth communications, but certain implementations also enable indoor precision tracking and radar sensing, on top of communications. The unique capabilities of UWB driving it into future markets include:

- High Spatial Capacity—UWB channelization and bandwidth per square meter are ultimately limited by the number of impulses that can be discerned in time over the propagation distance. For example, with 0.25 ns impulses the upper limit on pulse rate is 4 billion pulses per second that, because of low power, are confined to ranges up to tens of meters, thus providing exceptional spatial capacities.

- High Channel Capacity and Scalability—Scalability accommodates various channel profiles to harness the desired data rate given a channel impulse response. Phase coding of impulses can simultaneously integrate impulses to improve the energy per data bit and can rake energy from multipath-delayed signal replicas.

- Robust Multipath Performance—Multipath signal replicas can be raked-received for superior performance indoors. In the limit, the total impulse power on average propagates with an inverse square law just like free space propagation.

- Very Low Transmit Power—Sub-milliwatt power levels spread over several gigahertz of bandwidth means that the UWB signals will not cause interference to current users of the spectrum and also will generally be less susceptible to detection.

- High System Link Rate—Data rates can be from a high in the hundreds of Mbit/s down to hundreds of kbit/s. Given a fixed power level, the data rate may be traded off for additional range.

- Location Awareness and Tracking—Some implementations of UWB signaling inherently provide 3-D sensing and tracking at centimeter accuracies.

Advancements in UWB technology promise the opportunity of creating unique solutions meeting emerging market needs. There are essentially three basic market spaces in which UWB will play a role:

- Wireless Communications—As bandwidth available to users increases, applications will continue to evolve to fill the available bandwidth and to demand further increases. On top of this increasing demand for bandwidth, the increase in mobile telephony and travel has spurred demand for bandwidth mobility, implying wireless technology. Initial applications of UWB will evolve from the existing market needs for higher-speed data transmission, but demand for multimedia-capable wireless is already driving multiple initiatives in the wireless standards bodies. UWB solutions will emerge that are tailored for these applications because of the available high bandwidth. In particular, high-density multimedia applications, such as multimedia streaming in "hot spots" like airports or shopping centers or even in multidwelling units, will require bandwidths not currently enabled by continuous-wave "narrowband" technologies. The ability to tightly pack high-bandwidth UWB "cells" into these areas without degrading performance will further drive the development of UWB solutions. Full duplex and simplex radio systems using microwatts of power have already been demonstrated with data rates from 32 Kbit/s to hundreds of Mbit/s at ranges in the tens of meters in office environments.

- Precision Tracking—As the mobility of people and objects increases, up-to-date and precise information about their location becomes a relevant market need. Although GPS and some E911 technologies promise to deliver some level of accuracy outdoors, current indoor tracking technologies remain relatively scarce and have accuracies on the order of 3 to 10 m. UWB implementations are an adjunct to GPS and E911 that allow the precise determination of location and the tracking of moving objects within an indoor space to an accuracy of a few centimeters or less. This in turn enables the delivery of location-specific content and information to individuals on the move and the tracking of high-value assets for security and efficient utilization. Although this is an emerging market segment, the accuracy provided by UWB will accelerate market growth and the development of new applications in this area.

- Radar—Finally, UWB signals enable inexpensive high-definition radar. With the new radar capability created by the addition of UWB, the radar market will grow dramatically and radar will be used in areas currently unthinkable. Some of the key new radar applications in which UWB is likely to have a strong impact include automotive sensors, collision avoidance sensors, smart airbags, intelligent highway initiatives, personal security sensors, precision surveying, and through-wall public safety applications. Through-wall radar is already being tested to assist law enforcement and public safety personnel in clearing and securing buildings more quickly and with less risk by providing the capability to detect human presence and movement through walls. Radar-enhanced security domes based on precision radar have already demonstrated the capability to detect motion near protected areas, such as high-value assets, personnel, or restricted areas. The dome is software configurable to detect movement passing through the edge of the dome but can disregard movement within or beyond the dome edge.

## 15.6 Addressing the Wireless Spectrum Squeeze with UWB

UWB operates at ultra-low power, transmitting impulses over multiple gigahertz of bandwidth. Each pulse, or pulse sequence, is pseudorandomly modulated, thus appearing as "white noise" in the "noise floor" of other radio frequency devices. UWB operates with emission levels commensurate with common digital devices such laptop computers and pocket calculators. Today we have a "spectrum drought" in which there is a finite amount of available spectrum yet a rapidly increasing demand for spectrum to accommodate new commercial wireless services. The defense community continues to find itself defending its spectrum allocations from the competing demands of commercial users and even other government users. UWB exhibits incredible spectral efficiency that takes advantage of underutilized spectrum, effectively creating "new" spectrum for existing and future services by making productive use of what appears as the "noise floor" in conventional receivers. UWB technology represents a win-win innovation that makes available critical spectrum to government, public safety, and commercial users.

## 15.7 Conclusions

UWB capabilities and limitations have been extensively covered in the media [see, for example, 16–18. The best applications for UWB are for indoor use in high-clutter environments. UWB products for the commercial market will make use of recent technological advancements in receiver design and will transmit at very low power (microwatts). UWB technology enables not only communications devices but also positioning capabilities of exceptional performance. The fusion of positioning and data capabilities in a single technology opens the door to exciting and new technological developments.

## References

[1]   R. A. Scholtz and M. Z. Win, "Impulse Radio," *Proceedings of IEEE Personal, Indoor and Mobile Radio Communications—PIMRC 1997,* Helsinki, Finland.

[2]   M. Z. Win and R. A. Scholtz, "Impulse Radio: How it Works," *IEEE Communications Letters*, Vol. 2, No. 1, January 1998.

[3]   Time Domain Corporation, Huntsville, AL: http://www.timedomain.com.

[4]   Aether Wire & Location, Inc., Nicasio, CA: http://www.aetherwire.com.

[5]   K. Siwiak, "Ultra-Wide Band Radio: Introducing a New Technology," Invited Plenary Session Paper, *Proceedings of the IEEE Vehicular Technology Conference—2001,* Rhodes, Greece, May 2001.

[6]   M. Welborn, "System Considerations for Ultra-Wideband Wireless Networks", *Proceedings of IEEE Radio and Wireless Conference—RAWCON 2001*, Boston, MA, August 19 – 22, 2001, pp. 5–8.

[7]   H. G. Schantz and L. Fullerton, "The Diamond Dipole: a Gaussian Impulse Antenna", *IEEE APS Conference*, Boston, MA, July 2001.

[8]   A. J. Kerkhoff, "The Use of Genetic Algorithm Approach in the Design of Ultra-Wide Band Antennas," *Proceedings of IEEE Radio and Wireless Conference—RAWCON 2001*, Boston, MA, August 19–22, 2001.

[9]   K. Siwiak, T. M. Babij, and Z. Yang, "FDTD simulations of ultra-wideband impulse transmissions," *Proceedings of IEEE Radio and Wireless Conference—RAWCON 2001*, Boston, MA, August 19–22, 2001.

[10]  R. Hoctor, "Transmitted-reference, Delay-hopped Ultra-Wideband Communications," GE Corporate Research and Development, Forum on Ultra-Wide Band, October 11-12, 2001, Intel Corporation, Hillsboro, OR, http://www.ieee.or.com/IEEEProgramCommittee/uwb/uwb.html

[11]  K. Siwiak and A. Petroff, "A Path Link Model for Ultra-Wide Band Pulse Transmissions," *Proceedings of the IEEE Vehicular Technology Conference––2001*, Rhodes, Greece, May 2001.

[12]  D. Cassioli, M. Z. Win, and A. F. Molisch, "A Statistical Model for the UWB Indoor Channel," *Proceedings of the IEEE Vehicular Technology Conference––2001*, Rhodes, Greece, May 2001.

[13]  J. Foerster, "The Effects of Multipath Interference on UWB Performance in an Indoor Wireless Channel," *Proceedings of the IEEE Vehicular Technology Conference—2001*, Rhodes, Greece, May 2001.

[14]  K. Siwiak, "Impact of UWB Transmissions on a Generic Receiver," *Proceedings of the IEEE Vehicular Technology Conference—2001*, Rhodes, Greece, May 2001.

[15]  IEEE802 Standards meeting document, IEEE 802.15-00/195r8, October 2000, 00195r8P802-15_TG3-XtremeSpectrum-Multimedia-WPAN-PHY.ppt

[16]  W. Webb, "Ultra-Wideband: An Electronic Free Lunch?," *EDN Magazine*, December 21, 2000.

[17]  K. Siwiak and P. Withington II, "Ultra-Wideband Radios Set to Play," *EE Times*, February 26, 2001.

[18]  J. Cox, "Ultrafast Wireless Technology Set to Lift Off," *Network World*, Vol. 18, No. 35, August 27, 2001.

# Glossary

3G—Third Generation
5GWIAG—5-GHz Wireless Industry Advisory Group
AAA—Authentication, Authorization, and Accounting
AC—Access Controller
ACK—Acknowledgment
ACL—Asynchronous Connection Link
ADSL—Asymmetric DSL
AES—Advanced Encryption Standard
AFE—Analog Front End
AFH—Adaptive Frequency Hopping
AKA—Authentication and Key Agreement
AP—Access Point
API—Application Programming Interface
ARQ—Automatic Repeat Request
AS—Authentication Server
ATM—Asynchronous Transfer Mode
AWGN—Additive Wide Gaussian Noise
AWMA—Alternating Wireless Medium Access
BER—Bit Error Rate
BPSK—Binary Phase Shift Keying
BRAN—Broadband Radio Access Network
BSA—Basic Service Area
BSS—Basic Service Set
BSSID—BSS Identifier
BTS—Base Transceiver Station
CCA—Clear Channel Assessment
CCI—Co-Channel Interference
CCK—Complementary Code Keying
CDMA—Code Division Multiple Access
CDR—Call Detail Record
CEPT—European Conference of Postal and Telecommunications Administrations
CFP—Contention-Free Period
CHAP—Challenge Handshake Authentication Protocol
COFDM—Coded OFDM
CP—Contention Period

CSD—Circuit Switched Data
CSMA—Carrier Sense Multiple Access
CSMA/CA—CSMA with Collision Avoidance
CTS—Clear to Send
DA—Destination Address
dB—decibel
DBPSK—Differential Binary Phase Shift Keying
DCF—Distributed Coordination Function
DCS—Dynamic Channel Selection
DECT—Digital Enhanced Cordless Telecommunications
DFS—Dynamic Frequency Selection
DHCP—Dynamic Host Configuration Protocol
diffserv—Differentiated Services
DIFS—DCF Interframe Space
DLC—Data Link Control
DNS—Domain Name Service
DPSK—Differential Phase Shift Keying
DQPSK—Differential Quatemary Phase Shift Keying
DS—Distribution System
DSL—Digital Subscriber Line
DSSS—Direct Sequence Spread Spectrum
EAP—Extensible Authentication Protocol
EAPOL—EAP over LAN
ECP—European Common Proposals
EDGE —Enhanced Data Rate for Global Evolution
EESS—Earth Exploration Satellite Service
EIRP—Equivalent Isotropically Radiated Power
ERC—European Radiocommunications Committee
ESS—Extended Service Set
ESSID—ESS Identification
ETSI—European Telecommunications and Standard Institute
FBWA—Fixed Broadband Wireless Access
FCC—Federal Communications Commission
FDD—Frequency Division Duplex
FEC—Forward Error Correction
FFT—Fast Fourier Transform
FHSS—Frequency Hopping Spread Spectrum
FSS—Fixed Satellite Service
FTP—File Transfer Protocol
FWA—Fixed Wireless Access
GAP—General Access Profile
GFSK—Gaussian Frequency Shift Keying
GGSN—Gateway GPRS Support Node

GMSK—Gaussian Minimum Shift Keying
GPRS—General Packet Radio Service
GSM—Global System Mobile
GTP—GPRS Tunneling Protocol
GUI—Graphical User Interface
HCF—Hybrid Coordination Function
HCI—Host Controller Interface
HCR-TDD—High Chip Rate-TDD
HiperLAN—High-Performance Radio LAN
HLR—Home Location Register
HTML—Hyper-Text Markup Language
HTTP—Hyper-Text Transfer Protocol
IAPP—Inter-Access Point Protocol
IBSS—Independent Basic Service Set
ICI—Intercarrier Interference
IEEE—Institute of Electrical and Electronic Engineers
IETF—Internet Engineering Task Force
IFFT—Inverse FFT
IFS—Interframe Space
IMSI—International Mobile Subscriber Identity
Intserv—Integrated Services
IP—Internet Protocol
IPSec—IP Security
IR—Infrared
ISI—Intersymbol Interference
ISM—Industrial, Scientific, and Medical
ISO—International Standards Organization
ISP—Internet Service Provider
IT—Information Technology
ITU—International Telecommunications Union
ITU-R—ITU Radio Sector
ITU-T—ITU Telecommunications Sector
IV—Initialization Vector
JRG—Joint Rapporteurs Group
L2CAP—Logical Link Control and Adaptation Protocol
LAN—Local Area Network
LCR-TDD—Low Chip Rate-TDD
LLC—Logical Link Control
LM—Link Manager
LMP—LM Protocol
MAC—Medium Access Control
MAP—Mobile Application Part
MD—Message Digest

MIB—Management Information Base
MMAC—Multimedia Mobile Access Communication Systems
MPDU—MAC Protocol Data Unit
MSC—Mobile Switching Center
MSDU—MAC Service Data Unit
MSS—Mobile Satellite Service
MT—Mobile Terminal
NAAP—Network-Access Authentication and Accounting Protocol
NAI—Network Access Identifier
NAV—Network Allocation Vector
NGSO—Non-Geostationary Orbit
NIC—Network Interface Card
NSS—Network Subsystem
OBEX—Object Exchange
OCB—Output Control Block
OFDM—Orthogonal Frequency Division Multiplexing
OSI—Open System Interconnection
OSPF—Open Shortest Path Forwarding
OTP—One Time Password
OWL—Operator Wireless LAN
PAN—Personal Area Network
PAP—Password Authentication Protocol
P8CC—Packet Binary Convolutional Coding
PC—Personal Computer
PCF—Point Coordination Function
PCU—Packet Control Unit
PDA—Personal Digital Assistant
PER—Packet Error Rate
PHY—Physical
PIFS—PCF Interframe Space
PIN—Personal Identity Number
PLCP—Physical Layer Convergence Procedure
PMD—Physical Medium Dependent
PMP—Point to Multipoint
PN—Pseudonoise
PPDU—PLCP Protocol Data Unit
PPP—Point to Point
PSDU—PLCP Service Data Unit
PSK—Phase Shift Keying
QAM—Quadrature Amplitude Modulation
QoS—Quality of Service
QPSK—Quaternary Phase Shift Keying
RA—Receiver Address

RAN—Radio Access Network
RADIUS—Remote Access Dial-In User Service
RF—Radio Frequency
RNC—Radio Network Controller
RSSI—Receive Signal Strength Indicator
RTS—Request to Send
SA—Source Address
SCO—Synchronous Connection-Oriented
SCORT—Synchronous Connection-Oriented with Repeated Transmission
SDP—Service Discovery Protocol
SFD—Start Frame Delimiter
SGSN—Serving GPRS Support Node
SIFS—Short IFS
SIG—Special Interest Group
SIM—Security Identity Module
SIR—Signal-to-Interference Ratio
SKE—Shared Key Exchange
SLA—Service Level Agreement
SMS—Short Message Service
SNMP—Simple Network Management Protocol
SNR—Signal-to-Noise Ratio
SOHO—Small Office, Home Office
SRES—Signed Response
SRP—Secure Remote Password
SRS—Broadcasting Satellite Service (*Service de radiodiffusion par satellite*)
SSL—Secure Sockets Layer
STA—Station
TBTT—Target Beacon Transmit Time
TCP—Transmission Control Protocol
TCS—Telephony Control Service
TDD—Time Division Duplex
TDMA—Time Division Multiple Access
TLS—Transport Level Security
TPC—Transmit Power Control
TTLS—Tunneled TLS
UDP—User Datagram Protocol
UMTS—Universal Mobile Telecommunications System
U-NII—Unlicensed National Information Infrastructure
USIM—UMTS Subscriber Identity Module
UTRAN—UMTS Terrestrial Radio Access Network
UWB—Ultra Wide Band
VLR—Visiting Location Register
VPN—Virtual Private Network

WAE—Wireless Application Envirorment
WAN—Wide Area Network
WAP—Wireless Application Protocol
WBFH—Wide Band Frequency Hopping
WCDMA—Wide Band Code Division Multiple Access
WECA—Wireless Ethernet Compatibility Alliance
WEP—Wired Equivalent Privacy
Wi-Fi—Wireless Fidelity
WirelessHUIvIAN—Wireless High-Speed Unlicensed Metropolitan Area
    Networks
WISP—Wireless ISP
WRC—World Radiocommunication Conference

# Related Web Sites

IEEE 802 Standards (free download)
http://www.ieee802.org

IEEE 802.11 Working Group for Wireless Local Area Networks
http://www.ieee802.org/11

IEEE 802.15 Working Group for Wireless Personal Area Networks
http://grouper.ieee.org/groups/802/15/index.html

IEEE Wireless Standards Zone
http://standards.ieee.org/wireless

ETSI Standards
http://www.etsi.org/getastandard/honie.htm

Bluetooth
http://www.bluetooth.com

# About the Author

Benny Bing (*bennybing@ieee.org*) is a member of the research faculty with the Broadband Institute at the Georgia Institute of Technology, USA. He is the author of over 30 technical papers and 5 books, including *Wireless Local Area Networks,* which has been adopted by Cisco Systems worldwide. In addition, he is guest editor on Wireless LANs for the *IEEE Communications Magazine, IEEE Wireless Communications* (formerly *IEEE Personal Communications*), and the *IEEE Journal on Selected Areas in Communications.* He is a frequent lecturer on wireless LAN subjects, having recently conducted a customized course for Qualcomm Inc., in San Diego, California. Benny serves on the technical program committees of three major IEEE networking conferences and organizes the *International Conference on Wireless LANs and Home Networks* (www.icwllin.org). Recently, he was featured in the *MIT Technology Review* in a special issue on wired and wireless technologies. His current research interests include communication networks, protocol design, and queuing theory.

# Index

Printed in the United States
By Bookmasters